Polymers for Personal Care Products and Cosmetics

RSC Polymer Chemistry Series

Editor-in-Chief:
Professor Ben Zhong Tang, *The Hong Kong University of Science and Technology, Hong Kong, China*

Series Editors:
Professor Alaa S. Abd-El-Aziz, *University of Prince Edward Island, Canada*
Professor Stephen Craig, *Duke University, USA*
Professor Jianhua Dong, *National Natural Science Foundation of China, China*
Professor Toshio Masuda, *Shanghai University, China*
Professor Christoph Weder, *University of Fribourg, Switzerland*

Titles in the Series:
1: Renewable Resources for Functional Polymers and Biomaterials
2: Molecular Design and Applications of Photofunctional Polymers and Materials
3: Functional Polymers for Nanomedicine
4: Fundamentals of Controlled/Living Radical Polymerization
5: Healable Polymer Systems
6: Thiol-X Chemistries in Polymer and Materials Science
7: Natural Rubber Materials: Volume 1: Blends and IPNs
8: Natural Rubber Materials: Volume 2: Composites and Nanocomposites
9: Conjugated Polymers: A Practical Guide to Synthesis
10: Polymeric Materials with Antimicrobial Activity: From Synthesis to Applications
11: Phosphorus-Based Polymers: From Synthesis to Applications
12: Poly(lactic acid) Science and Technology: Processing, Properties, Additives and Applications
13: Cationic Polymers in Regenerative Medicine
14: Electrospinning: Principles, Practice and Possibilities
15: Glycopolymer Code: Synthesis of Glycopolymers and their Applications
16: Hyperbranched Polymers: Macromolecules in-between Deterministic Linear Chains and Dendrimer Structures
17: Polymer Photovoltaics: Materials, Physics, and Device Engineering

18: Electrical Memory Materials and Devices
19: Nitroxide Mediated Polymerization: From Fundamentals to Applications in Materials Science
20: Polymers for Personal Care Products and Cosmetics

How to obtain future titles on publication:
A standing order plan is available for this series. A standing order will bring delivery of each new volume immediately on publication.

For further information please contact:
Book Sales Department, Royal Society of Chemistry, Thomas Graham House, Science Park, Milton Road, Cambridge, CB4 0WF, UK
Telephone: +44 (0)1223 420066, Fax: +44 (0)1223 420247
Email: booksales@rsc.org
Visit our website at www.rsc.org/books

Polymers for Personal Care Products and Cosmetics

Edited by

Xian Jun Loh
Institute of Materials Research and Engineering, Singapore, Singapore
Email: lohxj@imre.a-star.edu.sg

ROYAL SOCIETY OF CHEMISTRY

THE QUEEN'S AWARDS FOR ENTERPRISE: INTERNATIONAL TRADE 2013

RSC Polymer Chemistry Series No. 20

Print ISBN: 978-1-78262-295-6
PDF eISBN: 978-1-78262-398-4
EPUB eISBN: 978-1-78262-800-2
ISSN: 2044-0790

A catalogue record for this book is available from the British Library

Published by The Royal Society of Chemistry,
Thomas Graham House, Science Park, Milton Road,
Cambridge CB4 0WF, UK

Registered Charity Number 207890

For further information see our web site at www.rsc.org

Printed in the United Kingdom by CPI Group (UK) Ltd, Croydon, CR0 4YY, UK

Preface

The research into polymeric materials for cosmetic and personal care applications is rapidly evolving and this book is intended to serve as a guide through the latest work, offering summaries of the current state-of-the-art cutting-edge research as well as extensive references to the latest breakthroughs. Each chapter carries detailed coverage on its specific topic and can be read on its own and the reader is advised to consult the references therein. The work behind the chapters in this book involved many hours of literature search and consolidation and the information is succinctly condensed into this volume. Readers can expect to be taken through the entire spectrum of materials development for cosmetic and personal care applications, from basic fundamental research, applied and platform technologies, and commercially viable applications. Looking ahead to the topics covered in the book, we begin by giving a broad overview of the market of polymers for personal care and cosmetics and covering some of the newest and most exciting trends in this field. Chapter 2 covers the use of natural polymers in personal care. Chapter 3 describes the recent developments in the development of acrylates for personal care, while Chapter 4 highlights the use of these polymers as rheological modifiers. Chapter 5 describes the growing area of polymers for antibacterial applications, which are important as preservatives. Chapter 6 talks about the use of polymers in the potentially exciting area of 3D printing and, through the use of stimuli-responsive polymers, the evolution of "4D printing." Chapter 7 touches on the use of nanoparticles in personal care and also covers the issues and debates related to their use in personal care. Chapter 8 describes the workhorse of personal care, silicones. Chapter 9 focuses on the new area of developing cyclodextrin-based materials. Chapter 10 describes the new area of thermogels and Chapter 11 introduces the use of a natural material, pectin, as a potential rheological modifier for personal

RSC Polymer Chemistry Series No. 20
Polymers for Personal Care Products and Cosmetics
Edited by Xian Jun Loh

Published by the Royal Society of Chemistry, www.rsc.org

care applications. Finally, Chapter 12 leaves some perspectives and thoughts on the future for the development of materials for this industry.

I would like to gratefully acknowledge the contributions of the authors who have embarked on this project together with us. I would also like to acknowledge the help of the friendly staff at the Royal Society of Chemistry for their patience with us at various stages of the project. I would like to express my deepest gratitude to Joo Gek Lim and Moi Joo Loh who have been actively involved in the language editing of this book and given their fervent support to this project. This book would not have been possible without their collective inputs and, indeed, the book is now much better because of their contributions. Finally, we hope that this volume will serve as an indispensable reference for students, researchers, academics and industrialists in the field of polymers for personal care research.

Xian Jun Loh

Contents

RSC Polymer Chemistry Series No. 20
Polymers for Personal Care Products and Cosmetics
Edited by Xian Jun Loh

Published by the Royal Society of Chemistry, www.rsc.org

CHAPTER 1

A Global Analysis of the Personal Care Market

CALLY OWH[a], PEI LIN CHEE[a], AND XIAN JUN LOH*[a,b]

[a]Institute of Materials Research and Engineering, A*STAR (Agency for Science, Technology and Research), 2 Fusionopolis Way, Innovis, #08-03, Singapore 138634, Singapore; [b]Department of Materials Science and Engineering, National University of Singapore, Singapore 117574, Singapore
*E-mail: lohxj@imre.a-star.edu.sg

1.1 Introduction

In keeping up with its evolution over the years, the scope of personal care products has been expanded to encompass products beyond conventional beauty and cosmetic merchandise. Following the increasing expectations of consumers with greater purchasing power than before, the range of personal care products has expanded to target healthy hair, nails and skin.

In fact, this increase in consumer purchasing power is but one of the many global trends, including globalization, technological advancement, greater consumer awareness and preferences for natural ingredients, that are contributing to create the favorable environment that is driving the growth of the personal care market in ASEAN.

As an effect of the aforementioned trend of globalization, more companies have been setting up their bases at strategic locations in order to take advantage of the benefits each location can offer (Figure 1.1).

RSC Polymer Chemistry Series No. 20
Polymers for Personal Care Products and Cosmetics
Edited by Xian Jun Loh

Published by the Royal Society of Chemistry, www.rsc.org

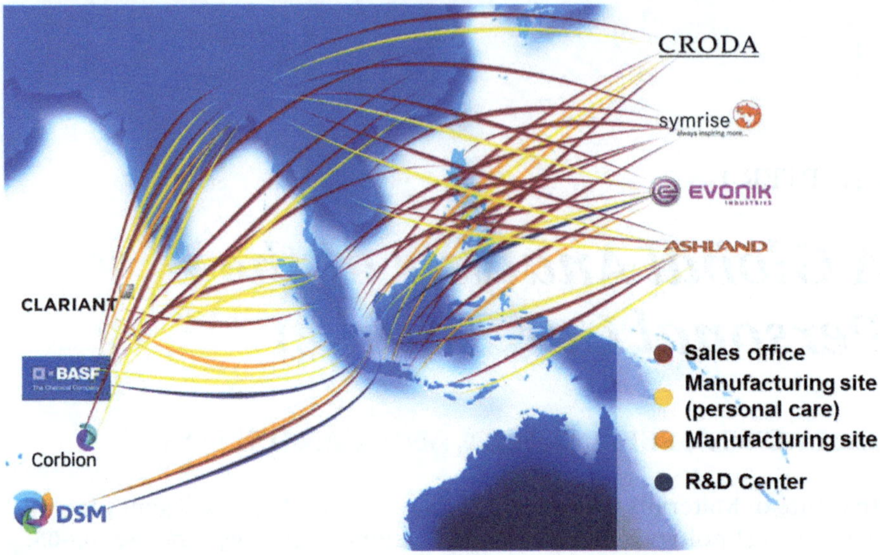

Figure 1.1 The establishment of production bases and innovation centers in ASEAN by established players.

This relocation enables them to be nearer to their potential markets, giving them better adaptability to respond to the changes in consumer tastes. Additionally, the production costs can be lowered in the less developed countries. As a result of this, numerous personal care product manufacturers and ingredients suppliers have been shifting their bases to emerging markets such as ASEAN, Latin America (LATAM), India and China. Although the original intention behind this shift of base was solely for the benefit of the companies, this relocation of bases has also brought about high economic growth and infrastructure development in emerging countries such as Indonesia and Thailand.

The current state of macroeconomic factors also creates a conducive environment for the growth of the market. Major ASEAN and LATAM economies are experiencing a strong growth in GDP, as well as a rise in incomes. This has resulted in progressively increasing living standards, which, together with rising consumer spending power, facilitates the aforementioned growth in the market. With the strong GDP growth and the onset of urbanization, the rise of the urban middle class population brings about an increased capacity for consumption, where such consumers can not only satisfy their necessities, but their desires as well. Among these desires, personal care grooming remains one of the closest to heart, with its importance further elevated by the propagation of beauty ideals through mainstream media, allowing for an observable growth in the cosmetic industry.

The regional age demographic in ASEAN is yet another contributor to the predicted rapid growth of the local personal care market. With a significant proportion of the population already in the 40–64 age group, and with a further expected growth in that population, the population

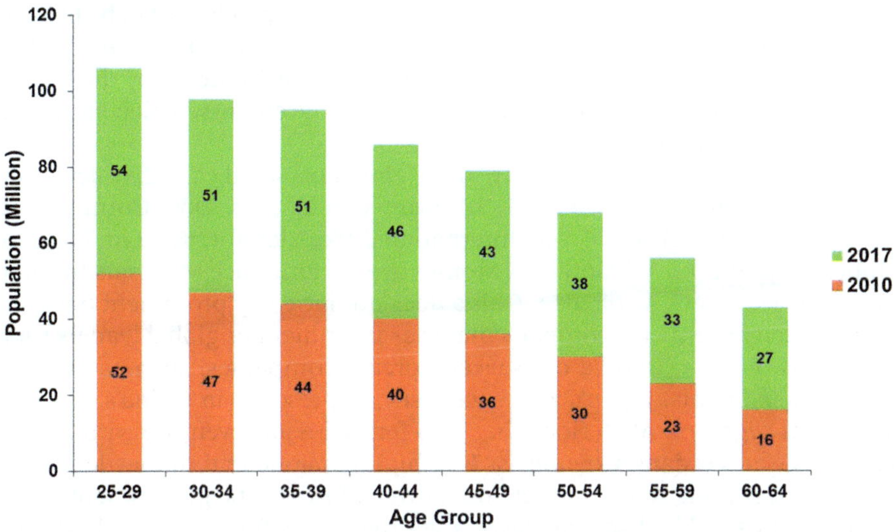

Figure 1.2 Population distribution in ASEAN for ages 25–64 years, 2010–2017.

distribution of ASEAN is indicative of a potential market in anti-ageing cosmetics (Figure 1.2).

Furthermore, although it has been observed that the conventional customers for anti-ageing products generally lie within what is considered the baby-boomers population, trends have now shifted with consumers opting to use anti-ageing products at an earlier age of 25 years. This shift is thought to be attributed to premature ageing as a result of lifestyle habits, such as the consumption of junk food, as well as the lack of exercise, driving forward the demand for anti-ageing products within the age group of 25–39.

Consequently, there has been an increased demand and consumption of cosmetic products, which are favored according to the active ingredient used. The rise in consumption of cosmetic products has been forecasted to continue for the next five years, potentially generating more wealth for the emerging markets than for the developed countries. Polymers are widely used in personal care and cosmetic products, such as emollients, conditioning polymers, humectants, emulsifiers, surfactants, thickeners, rheology modifiers, hair fixative polymers and opacifiers. According to market forecasts by Lucintel, the global polymer ingredients used in personal care products are expected to reach \$3.49 billion by 2017 (http://www.cosmeticsdesign-europe.com/Formulation-Science/Global-personal-care-polymers-market-predicted-to-reach-3.49bn-by-2017). Surfactants have the largest market share in terms of sales, while conditioning polymers are growing at a rapid rate. Large investments towards innovation and polymer development have been pumped in, which are driven by increasing consumer awareness and demand towards multi-functional and safe ingredients, as well as an environmental push for natural and green ingredients.

The personal care market is largely driven by skin care products, of which the consumers seek skin care products that are for anti-ageing, whitening and sensitive skin. The mega trends driving skin care products can be broadly classified into two categories: consumer-driven and environment-driven. Yet, the introduction of new ingredients to the industry is hindered by the ban on animal testing for cosmetic products and their ingredients, which started in the European Union from March 2009.

Leading global players operating in the global personal care chemicals and ingredients market include BASF SE, Evonik Industries, Dow Corning, Solvay S.A., Akzo Nobel N.V., Croda International Plc, Clariant, KCI Limited, J.M. Huber Corporation, Ashland Inc., Momentive Performance Materials Inc. and Wacker Chemie AG. As of 2011, China does not have any prominent personal care specialty chemicals company that is as prominent as global leaders Clariant, DSM, Evonik, Rhodia or Wacker. Foreign companies, such as France's Rhodia and Cognis, Japan's Kao have a strong presence in China's specialty surfactant segment. Akzo Nobel has also formed a joint venture with Zhangjiagang-based Feixiang Chemicals. In a highly fragmented domestic market which is beginning to consolidate under the efforts of the government, a few larger entities will be likely to play a larger role in the future of the global personal care chemicals and ingredients market, such as Guangzhou Tinci Materials, Sinopec, SinoChem, ChemChina, PetroChina and Shanghai Huayi.

1.2 Commercialization of Stimuli Responsive Polymers

Stimuli responsive polymers are polymers that adapt to environmental stimuli, such as temperature, pH, electrical signal, magnetic field, mechanical energy and ions, which give rise to a response such as a change in shape, permeability, phase, mechanical properties, optical properties and electrical properties. The response may be temporary or reversible; as such the original properties may be reverted to when the stimuli is removed or changed. Some of the polymers may respond to more than one stimulus simultaneously and in a predictive manner. Such stimuli responsive polymers are often referred to as smart materials, polymer chameleons and adaptive materials.

One of the first classes of smart materials that was discovered in the 1880s is electro-active polymers, which exhibit a change in size or shape when stimulated by an electric field. However, the breakthrough of such materials took close to 80 years when polyvinylidene fluoride (PVDF) was discovered in the late 1960s.

Increasingly, stimuli responsive polymers are used in more and more applications, such as drug delivery, diagnostics, tissue engineering and 'smart' optical systems, as well as biosensors, micro-electromechanical systems, coatings and textiles. The stimuli responsive polymers can be in the form of colloid, micelle, gel, capsules, emulsions, films and so on, depending on the use. However, much of the development work is still at laboratory scale. Application is limited as the mechanical and chemical stability of the polymers are low to perform its intended functionality, particularly in application sectors where high mechanical and chemical performance are required.

Spanish researchers at IKA-CIDETEC Research Centre developed a self-healing polymer that does not require external heat or light triggers. The material is based on bis(4-aminophenyl) disulfide as a cross-linker and aromatic disulfide metathesis for reversible bond formation and breakage that provide the self-healing behavior. However, commercialization of such material will require improvement in robustness. Raslabs, a company founded in 2003, developed electroactive polymer (EAP), of which one application is synthetic muscles.

1.3 Stimuli Responsive Polymers As a Personal Care Ingredient

The use of stimuli responsive materials in personal care ingredients has been highly researched as such materials are known to be able to encapsulate active ingredients for sustainable drug delivery. Water-soluble polymers and thickeners have been around for many years, and there is a growing trend for thermo-associative thickening at desired temperatures. There are many potentially "marketable" advantages of such materials in personal care products. Some are listed in Table 1.1.

1.4 Commercialization of Bio-Based Polymers

Bio-based polymers are starting to be maturing, with major investments, capacity scale-ups, numerous product launches, mergers and acquisitions (M&A) in a highly competitive market. The market drivers are the increase

Table 1.1 Examples of uses for different stimuli responsive polymers.

Stimuli	Examples of uses
pH-responsive	Encapsulation of actives or enzymes for detergents Food processing Nutricosmetics Encapsulation of active ingredients for targeted and/or sustainable delivery on skin Color-adaptive cosmetics Skin-pH balancing moisturizer
Temperature-responsive	Phase change of skin care lotion upon application Shape recovery for hair styling or mascara upon application As a shape memory mold for manufacturing of contact lenses Face mask that "wraps" the facial contour upon application
CO_2-responsive	Release of actives or O_2 on damaged skin for rejuvenation
Laser/light-responsive	Phase change of gel upon laser/light treatment, release of cooling agent for user comfort or added protection against heat
Electrical-responsive	Used with handheld devices for the release of sensitive actives

in environmental consciousness and advancement of technology. Emery Oleochemicals has grown its bio-polyol market (such as agricultural chemicals, bio-lubricants, and polymer additives) with the completion of a plant of 25 000 MT per year capacity in 2015. Cargill, BASF, Mitsui Chemicals, Itoh Oil Chemicals, Jayant Agro Organics are some companies in the bio-polyol industry. Ethanol producers like DuPont and Abengoa have shifted from fuels to biochemicals. DuPont has developed DuPont™ Sorona® EP Thermoplastic Polymer, which contains 37% plant-derived renewable material by weight and has a performance similar to conventional polybutylene terephthalate (PBT) plastics, typically used for automotive parts to furniture and mobile phone housings.

Bio-based polymers are applicable in a wide range of industries. Elevance and Stepan released a bio-based surfactant. Invista announced it was commercializing bio-based Spandex. Avantium are developing polymers with improved barrier properties, while Biosynthetic Technologies are developing improved performance bio-based lubricants.

Government and regulatory bodies play a significant role in pushing industry efforts in bio-based polymers. For example, since the Italian government passed a plastic bag ban in 2011, biodegradable bags became rampant in supermarkets across Italy. Cereplast's revenue increased five times in 2013 after the company received huge orders for resins that were used to manufacture biodegradable plastic bags for the supermarkets. There was also a ripple effect in the EU, where other countries like France have also started to adopt similar bans. However, due to a highly competitive market and high production costs, companies face tremendous challenges. Cereplast filed for voluntary petition for Chapter 11 bankruptcy in early 2014, and subsequently was acquired by Trellis Earth. With downstream integration offered by Trellis, it is expected that the chances of commercialization success will be high as Trellis converts Cereplast's resins into finished bioplastic products, which it can sell to its existing customers in applications like cutlery, bags, and containers. Several other M&As have also been observed to compete in the highly competitive market. Polylactic acid (PLA) producer DaniMer Scientific merged with polyhydroxyalkanoate (PHA) manufacturer Meredian to develop customized biopolymers for food packaging applications. Metabolix and other producers are pursuing various partnerships to blend their product with other biopolymers in an effort to improve performance and develop for specific applications.

1.5 Bio-Based Polymers As a Personal Care Ingredient

There are efforts from several companies to help reduce fossil fuel dependence by developing bio-based (also known as renewably-sourced) biopolymers. Commonly known biopolymers (including protein and carbohydrates) have been used extensively in today's cosmetics. Table 1.2 lists some of the

Table 1.2 Table of biopolymers and their use in personal care.

Biopolymers	Origin	Use in cosmetics	Biopolymer manufacturer	Products using the biopolymer
Xanthan gum	Fermentation of whey, corn, wheat, dairy, or soy by bacterium *Xanthomonas campestris*	Thickener, emulsifier, rheology modifier	Cargill, CP Kelco, Aurochemical, Shandong Deosen Corporation, ADM, Jungbunzlauer	Linage Body Moisturizer, Welenda's Calendula Lotion, Lisa Hopeman's Japanese Agarwood Cream
Agar	Red thalloid algae *Gelidium* and *Gracilaria*	Thickener, stabilizer, gelling agent,	American Agar Company	
Cellulose gum	Tree pulp and cotton linters	Thickener, emulsifier, film former	Danisco (Zhangjiagang) Textural Ingredients Co. Ltd., Ashland	Olay Regenerist's Advanced Anti-Aging Moisturize Night Resurfacing Elixir
Hydroxyethyl cellulose (HEC)	Tree pulp and cotton linters	Rheology modifier, thickener, stabilizer	Hercules Aqualon, Amerchol Corporation	Olay Regenerist's Advanced Anti-Aging Moisturize Night Resurfacing Elixir
Methyl cellulose (MC)	Tree pulp and cotton linters	Thickener, emulsifier	Jindi Co., Ltd., Qingdao Tianya Chemical	Fiama Diwills Shower Gel, MAC Green Gel Cleanser, Precision Gommage Microperle Eclat Maximum Radiance Exfoliating Gel, Glowfusion™ Micro-Tech™ Intuitive Active Bronzer
Chitin and chitosan	Shells of shrimp and other sea crustaceans	Film former, antibacterial active agent	Jinlong, Golden- Shell Biochemical	Jassen Cosmetics Skin Excel
Gum arabic	Arab gum tree	Emulsifier, thickener	Gum Arabic Processing Company, CNI (Nixera), Agriproducts, A&R (Alland & Robert)	Sukicolor
Hyaluronic acid	Membrane proteins, skin, cartilage, and the vitreous humour	Humectant	Contipro Group, Novozymes, TS Biotech, Dongying Foster Biological Engineering Co. Ltd., Ildong Pharmaceutical	Skin Medispa's Pure Vitality Nourishing Mask

(*continued*)

Table 1.2 (*continued*)

Biopolymers	Origin	Use in cosmetics	Biopolymer manufacturer	Products using the biopolymer
Alginates	Brown algae, seaweed	Thickener, stabilizer	FMC BioPolymer, Alginate Industries Ltd	Zhermack Inc's mask
Polyglutamic acid	Fermented soybean	Emulsifier, emollient, film former, antiwrinkle active agent	Shandong Freda Biotechnology	
Hydroxypropyl starch phosphate (HSP)	Starch	Thickener, stabilizer	National Starch	Vaseline, Kerastase
Carrageenan	Red seaweed	Rheology modifier, thickener, stabilizer, gelling agent, antibacterial active agent	FMC	
Dextrins	Starch	Active agent carriers	National Starch	Luminizing Moisture Tint by Jouer Cosmetics
Pectin	Citrus fruits and sugar beet	Surfactant, stabilizer, gelling agent		Lakme Peach Milk Moisturiser, Diapoly's Uroi- Bijin
Gellan Gum	Gram-negative bacteria *Sphingomonas elodea*	Stabilizer, gelling agent	CP Kelco	
Pullulan	Fermentation of starch by fungus *Aureobasidium pullulans*	Film former, antiwrinkle active agent		EuokoY-40 Blueprint Resculpting Cream, FANCL Cleansing Powder, Suhada Seikatsu Washing Powder

commonly used biopolymers, mainly polysaccharides, that are used in cosmetics. Honeywell developed the Asensa NSC and NCL range, based on a polylactic acid biopolymer, where Asensa NSC is used as an exfoliant for bath and shower products, and Asensa NCL is micronized for pressed powders and color cosmetic applications. Penford Products, a division of Penford Corporation, has been developing bioproducts since 1895, and has a tie up with Univar USA Inc. to introduce PenCare™ DP cationic polymers as conditioning polymers for hair and skin care products.

1.6 Personal Care Products with Active Ingredients

The concerns for health and wellness, coupled with greater consumer awareness, have also caused an increased demand for personal care products with active ingredients. These two factors exert a great influence over the market and could directly affect the popularity of the products. For example, consumers' awareness of the effect of cosmetic products on the environment has resulted in the development of environmentally friendly Cognis-BASF light emollients as a substitute for silicone products. In addition, there has been a greater demand for sunscreen as consumers become more well-informed of the dangers posed by UV light. Currently, there are two main types of sunscreen in the market: physical and chemical sunscreen. Generally, they differ in terms of their active ingredients and working mechanisms. Physical sunscreens contain nanoparticles of titanium dioxide and zinc oxide. These active ingredients will protect the skin from UV radiation mainly *via* scattering and reflection.[1] On the other hand, the active ingredients of chemical sunscreens include aminobenzoic acid derivatives, benzophenones, cinnamates, salicylates, avobenzone, ecamsule, ensulizole, bemotrizinol and bisoctrizole. These active ingredients of chemical sunscreen will protect the skin from UV radiation *via* absorption. The particles in physical sunscreens always cause an uncomfortable feeling on the skin, while chemical sunscreens are easier to use but those synthetic chemicals may lead to unexpected side effects on the skin.

Recently, natural sun blockers are receiving more and more attention as most of them possess good radiation protection functions as well as antioxidant properties. Lignin is a biomass from wood and rich in aromatic rings. During the derivation of lignin *via* polymerization of monolignols, there is loss of electronic conjugation of the vinyl group *para* to phenol, which causes the formation of UV chromophores at the sites of coupling.[2] As a result, lignin possesses the capability of UV/visible absorption.[3] Moreover, studies have proven that lignin has UV resistance properties, which allow it to be incorporated into a wide variety of materials for UV protection applications. Although lignin possesses UV resistance properties, investigations on the photostability of lignin-based sunscreens were carried out due to concern about its service lifespan as a UV blocker in the sunscreens. Surprisingly, recent results showed that both UVA and UVB absorbance increased when the lignin-based [1]sunscreens were exposed to UV radiation for two hours.[4]

Besides excellent UV protection properties, the free radical scavenging ability of its phenolic groups gives lignin outstanding antioxidant properties. Another advantage of lignin is that over 50 million tons are produced annually, which makes it an affordable and sustainable material to be used in sunscreen and as antioxidant actives in cosmetics.

Technological advancement is yet another driver of market growth. Along with higher standards of living and higher purchasing power, consumer expectations of product standards have also risen. It is now expected that the product goes beyond simply delivering its function to possessing multiple functions and high performance efficiency. In order to meet the expectations of the consumers and remain competitive, companies are now investing heavily in research and development (R&D), consequently facilitating progress in the area of active ingredients. For instance, Dow Personal Care had developed Opulyn PQG Opacifier, which overcame the prevalent issue of striking a balance between conditioning benefits and creamy texture. Besides providing a solution to the traditional formulation challenge, Opulyn PQG Opacifier offered conditioning cleansing products long term stability and it could be used together with cationic polymers.[5] TILAMAR is another product series which benefits from the progress of technology. Various polymeric hair care products were introduced ranging from deep conditioning, fast drying without stickiness, ultimate hold and weightless feel to cater for the different needs of the consumers.[6] Advancement in technology has also led to the new development of Plantasil Micro, a renewable substitute for silicone in the shampoo industry. Apart from its novelty as an oil-in-water microemulsion shampoo, the addition of anionic surfactant and PQ-10 allows Plantasil Micro to display better conditioning effects than conventional silicone shampoos.[7,8] These new developments in personal care products are made possible with the advancement in technology, which allows companies to keep in pace with the expectation of the consumers, hence driving the growth of the personal care market.

Consumer preference is another growth driver in the personal care industry. As more information on cosmetic products is available to the consumers through various channels (media coverage, retail promotions, brand marketing, celebrity endorsements), there has been a surge in the demand for natural products. This response is related to the concern about self-safety as well as the environmental impact the products can cause. Furthermore, natural ingredients are more readily available than before due to innovations and this has made them more affordable. As such, natural products become attractive alternatives to synthetic formulations, which motivates companies to look into natural sources of production, with standardized quality. An example of a natural ingredient used in body cleansing products is Euperlan GREEN, the first natural ethylene oxide and amine-free pearlizing dispersion developed by Cognis (now a part of BASF).

Apart from global trends, regional demographics and macroeconomic factors, market analyses conducted have revealed both short and long term impacts on the ASEAN personal care market caused by consumer, product innovation and business related trends. As of the year 2013, consumer-related areas are expected to see an increase in opportunities with regards to five key aspects, namely in anti-ageing ingredients, skin lightening, premium/prestige products, male grooming, as well as in the shift towards natural ingredients. In particular, the anti-ageing ingredient and skin lightening industries have been foreseen to be highly impacted by major opportunities in terms of both short and long term prospects. The premium/prestige product market has been predicted to face growth as well, with a transition from a medium short term impact to a high one in the long run. Male grooming, on the other hand, is expected to face a consistent medium opportunity impact on the market. Although a low impact has been predicted in the shift towards natural ingredients, analyses have shown a prospective progress in this market to a medium long term impact on the opening of opportunities. Product innovation and customization related areas have also been predicted to be impacted by product personalization and nutricosmetics opportunities. In this case, the former has a potential medium impact on the market in terms of both short and long term prospects, while the latter is predicted to have low impact in both cases. In business-related areas, industry consolidations are currently thought to have a temporary medium impact on the market, and a high long term impact.

However, despite the numerous factors driving the growth of the ASEAN market forward, there are also significant numbers of growth restraints acting in opposition to this progression. For one, the introduction of novel active ingredients in ASEAN is severely restricted due to the current state of unfavorable legislation. As of now, there has been no single, standardized, approach towards personal care active ingredient registration within the ASEAN regions, where the personal care product regulations simply share some common features with the cosmetic regulatory framework in Europe. These commonalities are inclusive of cosmetic labeling requirements, cosmetic product registration requirements, ingredient listing, cosmetic claim guidelines, as well as guidelines for cosmetic Good Manufacturing Practices (GMPs). As Article 4.1 of the European regulation positions that member states should adopt the Cosmetic Ingredient Listings of the Europe Cosmetic Drive 76/768/EEC and formulate ingredients included in the listing in their cosmetic end products, a restriction of novel active ingredients is imposed on the ASEAN market. This, however, can potentially be changed upon the setting up of the ASEAN Scientific Cosmetic Body (ASCB), which is expected to function in a similar manner to the Scientific Committee on Cosmetic Products and Non-Food Products (SCCNFP) in the European Union (EU), positioning itself as a reviewer of the ingredients list and technical and safety issues. With this, ASEAN integration by the year 2020 can be expected by personal care ingredient manufacturers, potentially bringing about a positive impact on the ASEAN market through a harmonization of standards across the different countries. There are, however, still concerns over possible similar procedural delays to those faced in Europe following REACH, which stem from an inability to reach a consensus over issues, and eventually resulting in the inhibition of innovation.

The growth of the ASEAN market is further suppressed by the intense competition posed by the Chinese market, which largely reduces the costs borne by global suppliers. As the prices of Chinese products are relatively low, a price pressure is created on the entire market. This has resulted in the necessity for many manufacturers to reduce their product prices so as to stay competitive, bringing down the profit margins for the affected. It has been observed that consumers are likely to purchase low-cost cosmetic products that are sold by trusted local participants or retailers. With this in mind, BASF has made the move to close its manufacturing plants located in Europe, marking a shift towards Asia in an attempt to beat the low costs of Chinese active ingredient manufacturers. Customers within the ASEAN cosmetic industry are also offered choices to substitute ingredients with low cost options from local suppliers. Thus, it is predicted that there will be an increase on the hold of the market by larger ingredient manufacturers in the long run, such that an increasing emphasis will be observed on quality from Tier I personal care products companies, followed by Tier II and Tier III personal care companies. Furthermore, top personal care companies are increasingly looking to leverage the R&D and technologies developed by prominent ingredient manufacturers through partnerships when it comes to active ingredients.

Counterfeit and smuggled personal care products in the ASEAN market are yet other factors detrimental to its growth. For example, there is a high number of fake personal care products sold in the Vietnam and Philippine markets. In fact, approximately 60% of the total numbers of personal care products that are sold are illicit, smuggled or counterfeit. With the evasion of sales tax and import duty brought about by smuggled counterfeit products, there has been the potential for competitive pricing, as well as higher dealer margin, for such markets, posing a strong competition towards the growing ASEAN market. There has also been the potential for deliberate and fraudulent mislabeling of counterfeit personal care products, allowing the illicit dealers to maximize profits through the selling of substandard products.

Finally, growth restraint in the present market has also stemmed from low market credibility, a resultant effect of a lack of clarity with regards to claims made about ingredient results. Despite the fact that validated claims of the performance of active ingredients in personal care products can, in fact, increase market credibility, there have been no official documents as of yet to act in support of these claims. For example, the ability of vitamins E-, A-, and alpha hydroxyl acid-(AHAs) formulated creams to aid in the effective prevention of stretch marks has been claimed, despite the lack of official proof. Thus, with manufacturers failing to provide a detailed and transparent account of the benefits and effects of personal care products formulated using their active ingredients, there has been an increased dependency of consumers on the marketing claims of the said products, resulting in their preference for lower-cost substitute products that can achieve similar benefits. Furthermore, with the misleading cosmetic claims made by companies, such as the use of the term "hypoallergenic", which is meaningless according to the Food and Drug Administration, consumers can also be misled into purchasing substandard products.

Despite the current market restraints, market evaluations have been indicative of an immense potential in the ASEAN personal care market. Global and macroeconomic trends, as well as suitable regional demographics contribute to an overall positive impact on its growth, allowing for a possible higher rate of growth than in mature markets (Figure 1.3).

As of 2013, ASEAN has already been holding more than a quarter of the percent share in the global market, beating out North America and LATAM (Figure 1.3). Its retail sales have been expected to grow to 360 billion USD by the year 2017 (Figure 1.4), and its worth has been predicted to reach 27 billion USD by 2020, anchoring it as a focused region in the area of personal care (Figure 1.5a).

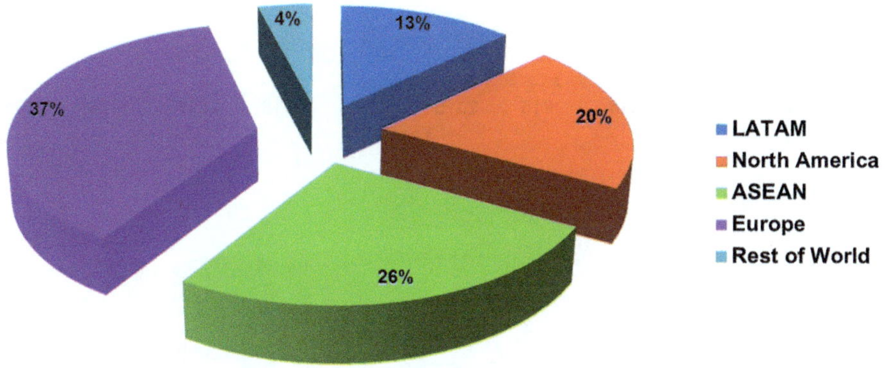

Figure 1.3 Percent share of personal care market by end users globally in 2013.

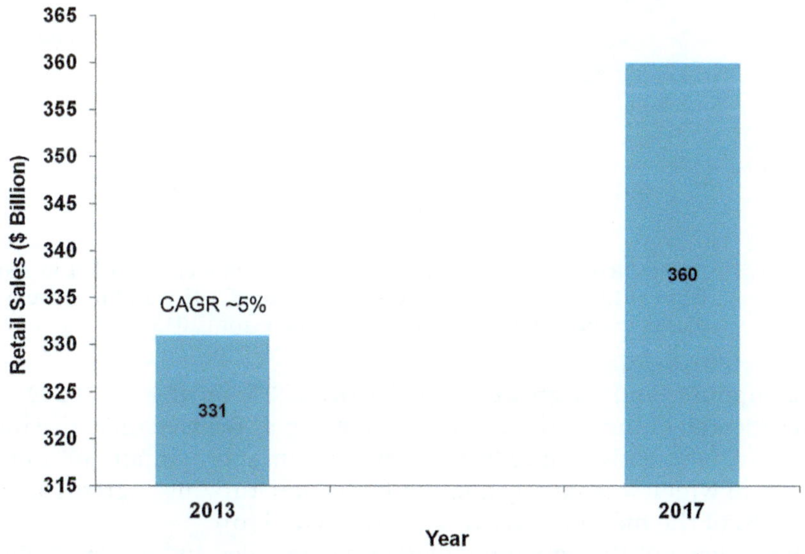

Figure 1.4 Predicted retail sales of the personal care market in ASEAN from 2013–2017.

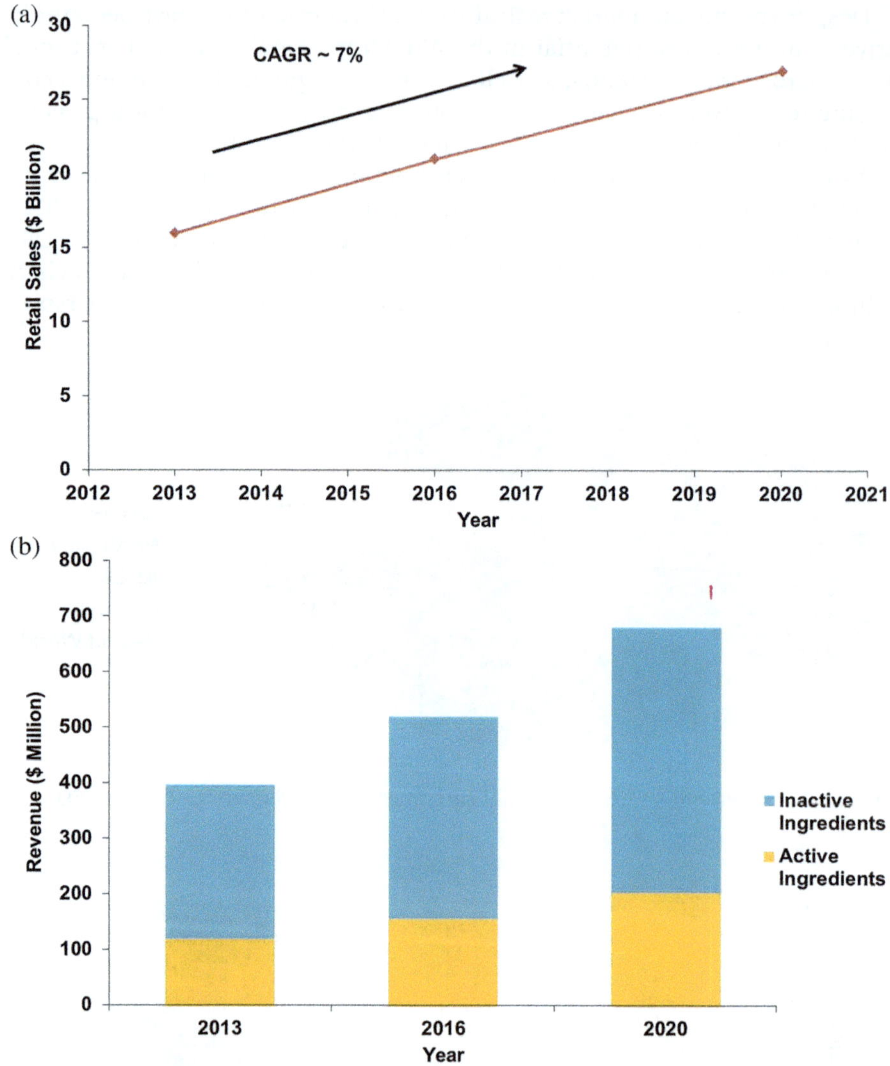

Figure 1.5 (a) Predicted retail sales trends of the personal care market in ASEAN from 2013–2020. (b) Expected significance of active and inactive ingredients in the ASEAN personal care market from 2013–2020.

A compound annual growth rate (CAGR) of 7% during 2013–2020 has been expected of the market, which is now largely focused on active ingredients. In fact, active ingredients make up an approximate 30% of the market, of which it is largely dominated by moisturizing ingredients that hold 85% of the market share (Figure 1.5(a) and (b)).

A breakdown of this emerging market has revealed Indonesia, Thailand, the Philippines, Malaysia and Vietnam to be the key players, with Indonesia and Thailand constituting an approximate 45% share of the entire market (Figure 1.6a). Apart from these, there has also been a recent rise in certain markets, notably Cambodia, Myanmar and Laos, marking their status as key emerging markets (Figure 1.6b).

Of the products on the market, which range from hair care to fragrances, the two products to have reached the highest end-use have been hair care products and skin care products, and these hold more than half of the market share together (Figure 1.7). It is interesting to note the variation in the emphasis on hair care based on cultural differences, with a higher hair care market share observed in countries with a non substantial female Muslim population who wear the Abaya for religious purposes.

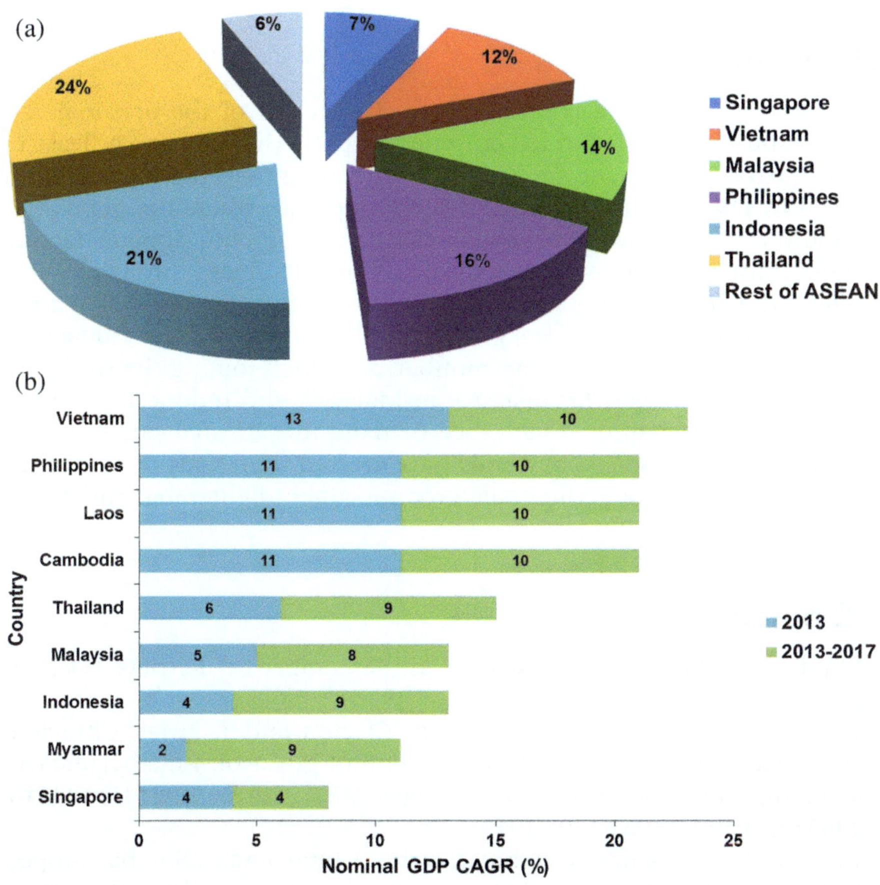

Figure 1.6 (a) Segmentation of the 2013 ASEAN personal care market by geography. (b) Predicted growth rates of the personal care market in Nominal GDP CAGR (%) of different regions in ASEAN from 2013–2017.

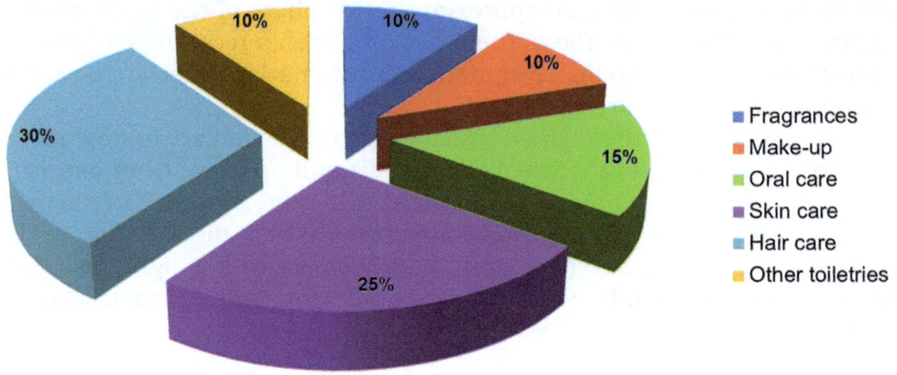

Figure 1.7 Percent share of personal care market by end users in 2013.

1.7 Conclusion

Many global trends have contributed to the growth of the personal care industry and are anticipated to continue powering the industry, at least up till 2017. Growing demand for moisturizing and anti-ageing ingredients is foreseen in Thailand, Indonesia and the Philippines where the population is rapidly expanding with favorable demographics and climatic factors. As more information is made available to the consumers through various channels, the demand for natural active ingredients in the multi-functional products is also believed to be a growth boosting factor for the industry. In addition, there are an increasing number of women joining the workforce in developing countries, for instance in Malaysia and Indonesia, and this would potentially lead to an increased demand for personal care products. Furthermore, all these global trends have created numerous opportunities for natural products targeting male grooming, skin lightening, anti-ageing and UV protection.

References

1. Physical vs. Chemical Sunscreen. (2012, September 11). Retrieved October 30, 2015.
2. J. Dean, P. Navotnaya, A. Parobek, R. Clayton and T. Zwier, Ultraviolet spectroscopy of fundamental lignin subunits: guaiacol, 4-methylguaiacol, syringol, and 4-methylsyringol, *J. Chem. Phys.*, 2013, **139(14)**, 144313–144316, DOI: 10.1063/1.4824019.
3. K. Chaochanchaikul, K. Jayaraman, V. Rosarpitak and N. Sombatsompop, Influence of lignin content on photodegradation in wood/HDPE composites under UV weathering, *Bioresources*, 2012, **7**(**1**), 38–55.
4. Y. Qian, X. Qiu and S. Zhu, Lignin: a nature-inspired sun blocker for broad-spectrum sunscreens, *Green Chem.*, 2014, **17**(**1**), 320–324, DOI: 10.1039/C4GC01333F.

5. http://www.sofw.com/index/sofw_de/sofw_de_produktneuheiten.html?naid=4341.
6. http://www.dsm.com/markets/personal-care/en_US/products/products-ranges/tilamar.html.
7. https://www.ulprospector.com/en/asia/PersonalCare/Detail/1133/91650/Plantasil-Micro.
8. https://www.ulprospector.com/en/na/PersonalCare/Detail/75/204181/Plantasil-Micro.

CHAPTER 2

Polymers for Personal Care – Natural Protein-Based Polymers

XIAO LU YIN[a] AND XIAN JUN LOH*[a,b]

[a]Institute of Materials Research and Engineering, A*STAR (Agency for Science, Technology and Research), 2 Fusionopolis Way, Innovis, #08-03, Singapore 138634, Singapore; [b]Department of Materials Science and Engineering, National University of Singapore, Singapore 117574, Singapore
*E-mail: lohxj@imre.a-star.edu.sg

2.1 What Is a Personal Care Product?

What first comes to mind when people hear the term "personal care product" is always a wide range of items that can commonly be found in the health and beauty sections of drug and department stores. The term "personal care product" is however not a defined legal term. Some products that are commonly referred to as "personal care products" are cosmetics, including a variety of products such as skin moisturizers, lipsticks, perfumes, facial and eye makeup preparations, shampoos, hair colors, toothpastes, deodorants and permanent waves. Others in contrast, are identified as drugs, such as skin protectants, mouthwashes marketed with therapeutic effects, antiperspirants and treatment for dandruff.

There are "personal care products" that meet the definitions of both cosmetics and drugs, and examples are products with two or more intended purposes. A shampoo is classified as a cosmetic because it cleanses the hair, an antidandruff treatment can be viewed as a drug. The combination of both

RSC Polymer Chemistry Series No. 20
Polymers for Personal Care Products and Cosmetics
Edited by Xian Jun Loh

Published by the Royal Society of Chemistry, www.rsc.org

treats dandruff and cleanses the hair at the same time, which most of the currently marketed shampoos can achieve. Other cosmetic/drug combinations include toothpastes with fluoride, deodorants that are also antiperspirants, UV-protective moisturizers and makeup. These products fulfil the requirement for both drugs and cosmetics.[1]

In general, "personal care products" can be viewed to comprise drugs (products intended to treat or prevent disease, or affect the structure or function of the body), cosmetics (products intended to cleanse or beautify) or a combination of both. This classification can also include other items like medical devices (such as certain hair removal and microdermabrasion devices), dietary supplements (such as vitamin or mineral tablets or capsules) or other consumer products (such as manicure sets).[2]

2.2 Personal Care Products—Market Overview

The global beauty market can be divided into a few major business segments. According to Statista, the five major components are: skincare, haircare, color (make-up), fragrances and toiletries.[2] Consumers' needs and expectations with regard to cosmetics are satisfied within these complementary segments. Skincare products made up more than 30% of global market share, followed by haircare with more than 20% market share, make-up, perfumes and toiletries and deodorants. Oral cosmetics have over 1% market share (Figure 2.1).

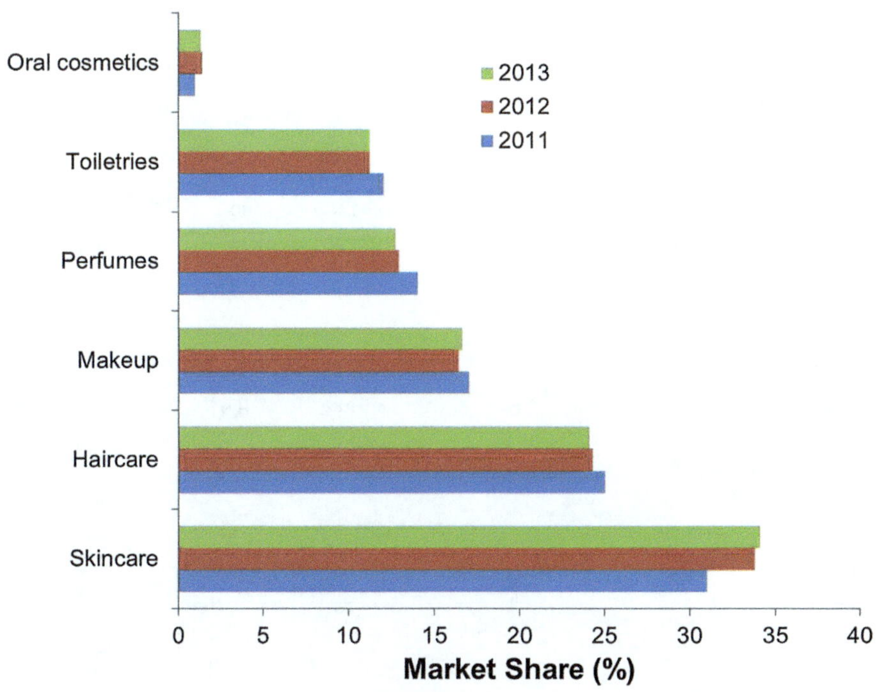

Figure 2.1 Market share breakdown of personal care products.

2.3 The Fundamentals: Amino Acids

The basic structural units of proteins are amino acids. Amino acids can be classified according to the value of their isoelectric point, structure and side chain polarity.

Examples of typical polar amino acids are shown in Figure 2.2.

Examples of typical non-polar amino acids are shown in Figure 2.3.

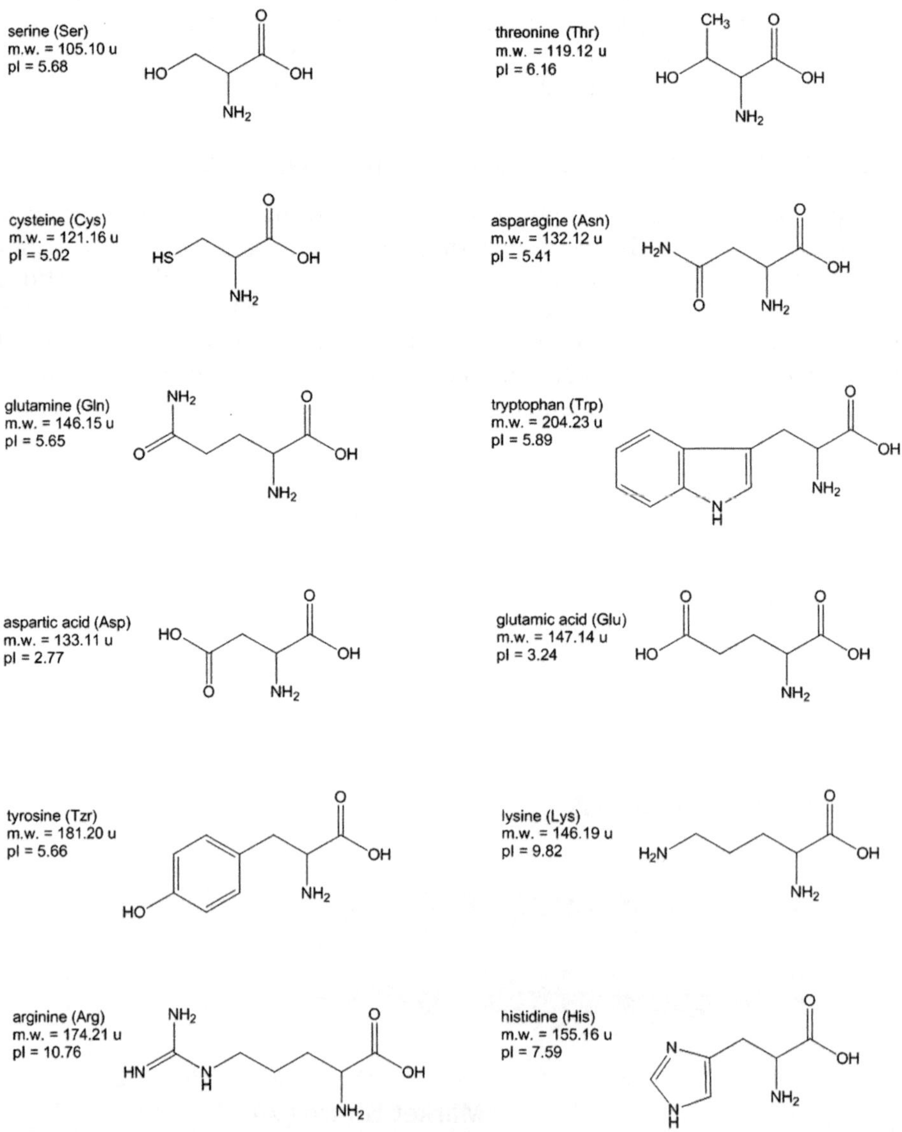

Figure 2.2 Types of polar amino acids.

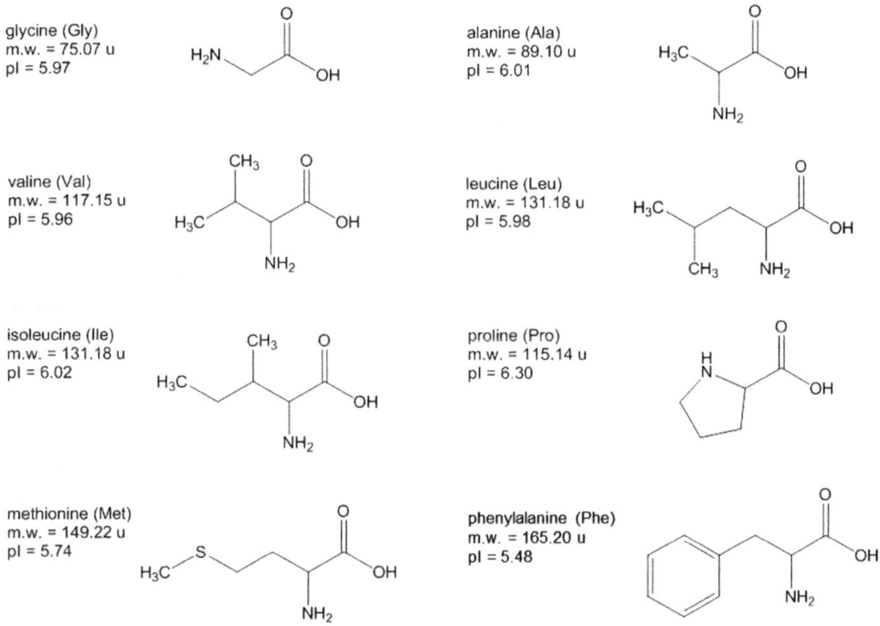

Figure 2.3 Types of non-polar amino acids.

Another way of classification is based on the source of the amino acids. Those amino acids synthesized in warm-blooded organisms are referred to as endogenous amino acids. Others are consequently labelled as exogenous amino acids.

2.3.1 Sources and Content of Amino Acids

The quantity and types of amino acid in a protein are determined by its genetic code and its source. Hydrolysates that are retrieved or derived from different proteins such as keratin, collagen, wheat and corn or soy gluten have similar amino acid content, showing that the extracts from the different sources of proteins can be used for different applications. The advancement of processing technologies has enabled the production of cosmetic raw materials from plant-based hydrolysates that have physical and chemical properties that are similar to animal-derived hydrolyzed proteins such as collagen. However, this approach is not feasible in the case of proteins with very special characteristic amino acid content. For example, elastin is a protein based on non-polar amino acids (up to 80%). Most protein-derived cosmetic raw materials are processed and obtained from simple globular and fibrous proteins that are found in animals (mammal, fish) or plant sources. Proteins that are obtained from lower organisms such as algae and fungi are less frequently used. Conjugated proteins such as proteoglycans and nucleo-proteins, which are extracted from animal or plant tissues, are seldom employed in cosmetic formulation. The most important criteria that determine the selection of the raw materials used as protein sources are costs, market value and availability.

2.3.2 Plant Source Proteins

With the emergence and increased lobbying of animal rights advocates and the fear of mad cow disease (Bovine spongiform encephalopathy), cosmetic raw materials produced from plant sources have started to replace those produced from animal sources and hydrolyzed implant proteins are not only used in the food industry.[3] However, implant proteins and their hydrolyzed product are different from animal proteins. The major function of plant proteins is to provide a source of nutrition. Therefore plant proteins have different characteristic amino acids and are generally stored in seeds. Plant proteins, for example gluten, are predominantly produced from wheat and corn grain (up to 80% concentration), soy and rice (up to 90% concentration) as well as defatted seeds. Potatoes are an alternative source of these proteins. The use of plant proteins allows manufacturers to label formulations as "possessing green ingredients" and provides consumers with the impression of using an environmentally friendly and safe product.[4]

Most plant-based proteins on the market are in the form of water soluble hydrolysates. They are recognized as solutions with a light color and with a cosmetic odor. Hydrolyzed vegetable protein is moisture-retentive and can be applied as a protective film-former. Plant proteins have no hydroxyproline, which is mainly found in proteins from animal sources such as collagen and elastin. In addition, some of the plant-derived proteins have eight exogenous amino acids which is different from tryptophan (Trp)-poor collagen-modified proteins. From a nutritional point of view, amino acid content in the plant proteins is not as balanced as animal proteins. For example, cereal proteins including wheat, corn and oats do have a high concentration of sulfur-containing amino acids, but lack lysine, threonine and tryptophan in the case of corn. Hydrolyzed oat protein has good emollient properties and its effect for hair conditioning is dramatic. Some studies suggest that oat protein can penetrate the hair shaft, protect damaged hair and maintain its structure with no buildup over time. Reduced irritation and inflammation as well as protection against damaging free radicals can be expected from lubricating and moisturizing agents from oat proteins.[5]

Proteins derived from beans and peas have limited amounts of methionine and threonine, but have a high content of lysine. Soy proteins have the highest nutritional and cosmetic values, with a very high content (up to 90%) of reserve globulins (glycinins and conglycinins). From an amino acid point of view, soy-derived proteins contain glutamine (20%) and asparagine (12%), arginine (9%) and leucine (8%), but contain very limited methionine, cysteine and threonine. Hydrolyzed soybean extract has served as an alternative to mammal-derived protein extracts in products designed to counteract the signs of biological and UV-induced premature skin aging. Products with high concentrations of glycoprotein and polysaccharides from soy flour have been employed in skin care products where skin firming and elastin protection is desirable.[6–8]

Proteins in wheat gluten are also widely employed as raw materials for cosmetic formulations. These kinds of proteins usually consist of two components, gliadin and gluten. Gliadin has a low molecular weight while gluten has a high molecular weight. Wheat gluten has low water solubility, mainly due to its high molecular weight and high amount of disulfide bridges. Its disulfide bond forming ability can be utilized for cosmetic effectiveness. From an amino acid compositional point of view, wheat gluten contains glutamines (33–40%) and prolines (13–15%). Two percent of the cysteine residue is in gliadin and 1.5% for gluten. Hydrolyzed wheat protein has extraordinary conditioning and antistatic properties for hair and skin, which have been extensively used in the formulation of conditioners, shampoo, cleansers, and skin care products.[9] Proteins derived from rice and beans have a significant amount of tyrosine. This amino acid is important in the formation of melanogenesis in which tyrosine is converted to tyrosinase, an enzyme that is responsible for the production of melanin. Tyrosine applied externally is able to hasten the process of melanin production and is therefore often used in tan acceleration cosmetic products. Oil plant protein hydrolysates that are derived from sunflower seeds and nuts can also be found on the market. The most popular and most used proteins are however derived from sweet almonds. Proteins obtained from cotton, lupine seeds and broad beans are less frequently employed despite the fact that the amino acid content is very similar to those derived from soy and wheat. The cosmetic effectiveness of these two types of proteins are also similar. Hydrolyzed sweet almond protein extract is used for its soothing and insulating properties and has been an ingredient where anti-irritation properties are desired.[10] It also has hydration and protection properties. Almond extract can also be used for reconstruction and conditioning of hair due to its high affinity to keratin in hair. The material consists of glucosides and peptides with branching and the film formed on hair improves its structural integrity and adds shine to hair fibers.[11]

The amino acid compositions in mg g^{-1} of commonly used plant proteins are shown in Table 2.1.

2.3.3 Animal Source Proteins

The first type of animal proteins used in cosmetic products is scleroproteins which can be found in animal tissue. Scleroproteins or fibrous proteins constitute one of the four main types of proteins (alongside globular, disordered and membrane proteins). There are many scleroprotein superfamilies including keratin, collagen, elastin and fibroin. They are used because of their availability, low production and process cost, high purity and similarity to human tissue. Proteins and hydrolyzed protein products derived from animals are the most popular and most widely used raw materials for cosmetic production. They have a wide range of molecular weights and are produced

Table 2.1 Amino acid composition of proteins and hydrolyzed proteins obtained from plants.

	Amino acid content of proteins [%]					Amino acid content of hydrolyzed proteins [mg g^{-1}]				
Amino acid	wheat	rice	soy	pea	almonds	pea	corn	soy	corn	potato
Alanine	2.7	5.0	4.2	5.6	4.2	3.68	2.32	0.74	6.22	2.22
Arginine	3.2	9.0	7.8	11.2	11.1	7.24	2.92	1.46	2.01	1.80
Asparaginine	3.3	8.3	12.2	14.4	9.9	9.76	3.25	2.28	3.28	6.01
Cysteine	2.3	1.4	2.3	—	1.6	0.74	1.08	0.26	1.14	0.30
Glutamine	40.2	20.9	19.5	23.2	31.9	16.40	33.20	4.16	9.05	5.67
Glycine	3.4	3.7	4.3	4.8	5.6	3.56	6.14	5.04	2.59	3.65
Histidine	2.3	1.5	1.4	2.4	2.3	2.19	1.55	0.54	1.21	<0.70
Isoleucine	11.9	13.4	1.0	9.6	9.6	3.77	2.71	0.80	1.83	2.08
Leucine	—	—	7.8	—	—	6.96	5.85	1.26	5.72	4.16
Lysine	1.5	4.9	9.5	7.2	3.2	6.24	1.62	0.54	1.55	2.47
Methionine	1.7	2.4	1.5	1.6	0.3	0.81	1.15	0.26	0.83	0.74
Phenylalanine	5.6	6.3	5.5	5.6	5.1	4.23	4.51	0.78	0.30	1.94
Proline	14.2	3.7	5.5	6.4	3.7	4.06	12.00	1.12	5.42	2.38
Hydroxyproline	—	—	—	—	—	—	—	—	—	—
Serine	5.3	5.0	4.8	11.2	3.1	3.03	4.64	1.12	2.74	1.91
Threonine	2.7	3.8	3.6	4.0	2.5	2.42	0.74	0.74	2.18	2.02
Tryptophan	—	1.3	—	—	—	0.42	0.09	0.24	<0.10	<0.10
Tyrosine	3.9	5.6	4.5	4.0	1.9	2.72	0.51	0.24	<0.10	<0.10
Valine	4.6	6.4	4.9	0.1	3.9	4.05	3.08	0.84	2.71	2.67

in the form of solutions with varying concentrations. Animal proteins still cannot be totally replaced or substituted by implant proteins due to their specific properties. Table 2.2 summarizes amino acid content compositions in the most widely used animal-derived proteins.

2.3.3.1 Collagen and Elastin

Collagen and elastin are naturally occurring proteins found in the skin dermis that are produced by fibroblasts, which are the major cells of the dermis. Collagen and elastin, together with some other proteins, form the extracellular matrix, which is the structural support that provides the strength, elasticity and durability of the skin. Collagen is the most abundant protein found in the skin, accounting for 70% of all the protein content. It maintains skin tone and elasticity. Elastin fibers and collagen fibers interlace and give the skin its elasticity. Collagen production slows down and loses its flexibility as people age and elastin fibers become thick and brittle. As the protein fiber diameters decrease, the skin becomes thinner. This whole process is externally displayed as wrinkles and dry skin. Collagen is the most often used animal protein. It is a structural protein and can be found in tissues like tendons, vascular walls and cartilages. It is the dominant component of the skin with more than half of all the collagen in the body being secreted in the skin (type I for 85–90%, type III for 8–11%). Collagen is secreted by fibroblasts that utilize pro-collagen units in the cytoplasm (type III collagen). The collagen formation process is summarized here: procollagen forms helical

Table 2.2 Amino acid composition of animal-derived proteins.

Amino acid	Collagen	Elastin	Hair keratin	Milk albumins	Silk	Casein
Alanine	9.8	16.6	4.5	7.3	26.8	3.1
Arginine	7.8	1.9	9.1	3.3	1.7	3.8
Asparaginine	4.9	2.4	5.8	10.8	5.6	7.1
Cysteine	—	—	16.3	1.3	0.1	0.4
Glutamine	9.6	4.4	12.7	15.0	4.4	22.0
Glycine	28.5	20.8	4.2	4.8	32.8	2.0
Histidine	0.6	0.2	0.8	2.2	1.1	2.9
Isoleucine	4.4	10.8	4.0	Trace	0.8	5.4
Leucine	—	—	6.4	12.1	1.2	9.5
Lysine	3.6	1.4	2.7	3.7	1.7	8.1
Hydroxylysine	1.0	—	—	—	—	—
Methionine	—	0.1	0.8	Trace	0.2	2.8
Phenylalanine	2.1	6.6	2.7	6.5	0.9	5.2
Proline	12.9	14.2	4.3	8.0	1.2	11.6
Hydroxyproline	10.6	—	—	—	—	—
Serine	3.2	1.6	7.6	7.2	11.6	6.0
Threonine	1.7	1.5	7.7	7.8	2.1	4.7
Tryptophan	—	—	—	—	—	1.7
Tyrosine	0.5	1.6	3.0	Trace	1.0	5.8
Valine	2.2	11.9	5.8	9.3	2.1	6.7

structures tipped with non-helical telopeptides. Hydroxylation of the built-in peptide chain with proline residues is an important stage in the formation of procollagen. This process is aided by proline hydroxylases, which are activated by ascorbic acid. Telopeptides of procollagen have a small amount of proline and hydroxyproline. They also have a large amount of tyrosine and cysteine that forms disulfide connections. Upon procollagen secretion, proteases remove telopeptides forming a triple twisted polypeptide strand. The helix is maintained and stabilized by the hydrogen bonding formed between hydroxyproline subunits and peptide bonds in each chain. The breaking down of the hydrogen bonding results in denatured collagen, which is known as gelatine. Collagen has high concentrations of glycine and alanine. Glycine appears at three amino acid intervals in the main chain. The structure of collagen can be summarized as $(\text{Gly–X–Y})_n$. The high content of proline and hydroxyproline can bind to water.

Most collagen used to be derived from calf connecting tissues, however it is currently excluded from cosmetic usage due to the emergence of mad cow disease.

Now collagen is obtained from marine creatures like fish skin as well as from poultry.

Collagen obtained from different species and different cells may vary in structure, composition and amino acid distribution, which results in different properties. Tables 2.3 and 2.4 present these differences in collagen derived from different sources.

Besides collagen, elastin is also widely used. Elastin is a highly elastic and cross-linked protein with a high amount of hydrophobic amino acids. The concentration of non-polar amino acids such as isoleucine is about 27%. Elastin is also a structural element that is responsible for the elasticity of tendons, ligaments and vascular walls and is another major component in

Table 2.3 Content of selected amino acids in collagen obtained from different sources (/1000 amino acids).

Amino acids	Poultry	Bovine (1)	Bovine (2)	Fish
Glycine	343.0	336.0	310.0	340.0
Hydroxyproline	105.0	83.0	106.0	70–84
Proline	119.0	122.0	145.0	109–151
Tyrosine	0.9	2.6	1.8	7.0

Table 2.4 Selected physicochemical properties of collagen obtained from different sources.

	Poultry	Bovine	Fish
Denaturation temperature (°C)	41–43	38	Max. 33
pH	4.0–4.5	2.3–4.0	3.0–4.5
Contamination content (%)	<0.3	<5	<1
Clarity of collagen solution	Transparent	Turbid	Turbid
Solubility	Good	Poor	Mean

connective tissue, besides collagen. It also has high amounts of hydroxyproline and glycine with many repetitions of the amino acid sequence – valine–proline–glycine–valine. The elasticity and durability of elastin can be attributed to the high concentration of desmosine that is responsible for cross-linking after the basic structure of elastin (tropoelastine) is synthesized. Isodesmosines are also present in elastin and also participate in the formation of cross-links (Figure 2.4).

As cosmetics are not used to treat diseased skin or skin wounds, they are not subject to the stringent FDA regulations imposed on drugs. Huc *et al.* showed that proteins used in cosmetics such as skin creams do not penetrate the skin barrier (*stratum corneum*).[12,13] Proteins with a large molecular size can form a continuous protective layer over the surface of compromised skin or hair without reducing the effect of cleansing or foaming activity of washing preparations.[12] The effectiveness of the protective proteins is evaluated by the proteins' absorption capacity into substrates leading to "substantivity."[12] This term is defined as the amount of protein or peptide which, after an extended exposure time, is resistant to extraction by water because of weaker or stronger linkages with the surface of the skin or hair. From this definition, hydrolyzed collagen or elastin with an appropriate molecular weight shows good activity and substantivity. Collagen is able to improve skin hydration and accelerates skin smoothing due to its ability to bind water. It possesses both skin-moisturizing and film-forming abilities. Its extraordinary smoothing and hydrating properties make collagen useful in skin- and hair-care applications. Hydrolyzed collagen from multiple animal sources has been employed in skin regenerating products that aim at reducing skin dryness, help mature skin and reduce damage to the skin caused by detrimental effects from the environment, such as UV exposure. Collagen with modifications has also been employed in shampoo and conditioners. Collagen with

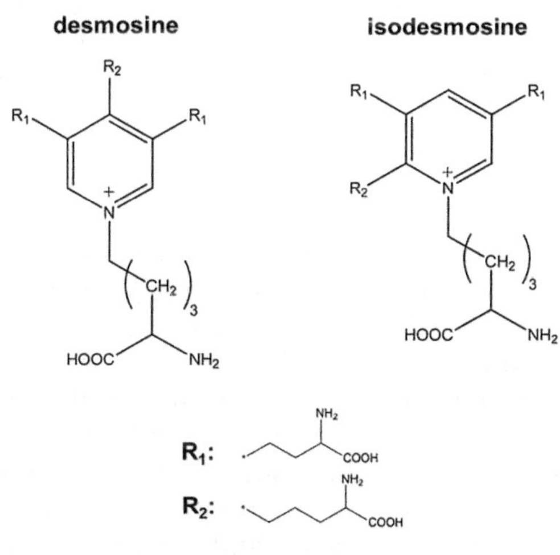

Figure 2.4 Structures of desmosine and isodesmosine.

quaternary ammonium salts incorporated has shown improved solubility. Human collagen that is similar to that in the skin is injected to reduce facial wrinkles. Hydrolyzed elastin can provide a natural protein layer of protection against environmental factors because of its high skin and hair affinity. Elastin has been added to skin-care formulations and has shown an effect in tightening mature skin and promoting regeneration of new elastic fibers.

Collagen and elastin have been extracted from fish. Specialized proteins produced from marine creatures have also been used. One example is the hydrolyzed conchiolin marine protein, produced from pearl oyster. It has been demonstrated to improve hair health. The protein can adsorb on hair cuticles, making the hair look smooth and shiny.[14]

2.3.3.2 Keratin

Keratin is found as a component in the epithelial cells and skin. It can be extracted from wool, feathers and hair. Keratin has two different structural components, with one being helical and the other non-helical, and these are produced from polypeptide chains interconnected by disulfide bonds. There are two types of keratin: α-keratins with high concentrations of cysteine and β-keratins which have high concentrations of glycine and tyrosine. Non-helical components produce the amorphous matrix in which α-keratin sections are displaced. Cysteine is the amino acid responsible for cross-linking formation. Keratin is not soluble in water due to its complex structure and hydrophobic amino acids. It is also the most abundant protein found in hair. Sulfur linking between keratin proteins forms disulfide chemical bonds with high strength and it is therefore difficult to break these bonds. The linking of the keratins determines the durability and resistance of hair fibers against environmental degradation. Hair perming uses alkaline solutions and reducing agents that break the disulfide bonds in the keratin proteins and this will result in hair damage with repeated treatment. Hydrolyzed keratin however, can penetrate the hair shaft to repair and rebuild damaged hair while its smoothness has been shown to be improved by cationic keratin containing cysteine.[15,16]

2.3.3.3 Silk

Silk is a continuous strand of two filaments joined together, producing the cocoon of the silkworm, and is increasingly being employed in hair and skin-care applications.[17] There are two types of proteins in it, fibroin and sericin, which contain many amino acids that are also constituents of the skin and hair. Proteins in silk are also compatible with a wide range of cosmetic ingredients, including ionic and non-ionic substances. One particular protein constituent of silk is soluble, has a high molecular weight and has been shown to bind to skin keratin and water, thereby forming a protective film. Similar to collagen and elastin, silk proteins also have smoothing, film-forming and hydrating properties that make them useful in a wide range of scalp, skin and hair-care products. They can increase gloss and improve manageability and texture of the hair. Functional hybrid polymers made of hydrolyzed silk protein with an alkyl chain and a silicone chain have been used as moisturizers, skin and hair protectors and antidiscoloring agents for hair.[18,19] The double strand of fibroin is held together by sericin, which is also known as silk gum and was demonstrated to have a shielding effect against UV radiation exposure. It gives protection against sunlight, chlorine, seawater and chemical treatments. Fibroin accounts for 70 to 80% and serine consists 20 to 30% of the total cocoon weight. In sericin, about 80% of the amino acid has a hydrophilic lateral group, about one third of which is serine, which has a water absorption capability 50 times higher than that of glycine. Sericin, therefore, has excellent moisturizing properties. It can form a protective film on the surface of skin and hair and preserve water content. Skin can be soft and smooth, hair can be soft and flexible upon application of sericin.[20]

1. Sericin is excellent at absorbing and releasing moisture. Sericin gives a gentle and smooth feeling on the skin. A silk-like fresh feeling is also provided by reducing sweat and wetness.[21]
2. Sericin has an amino acid structure similar to the natural moisturizing factor which is a natural component of cosmetic make up. Serine is the amino acid responsible for the moisturization. It does not stick to the skin and gives a gentle and soft touch even after repeated washing.[21]
3. Sericin reduces lipid peroxidation and prevents tyrosinase activity *in vitro* and thus contributes its antioxidant activities to hydroxyl group chelation with a small amount of copper and ions.[21] Tyrosinase is the enzyme responsible for production of skin melanin.[22]
4. Sericin also has an inhibitory effect on UV radiation induced acute damage by reducing oxidative stress in the skin of hairless mice.[21]

Sericin has an amorphous random coil and β-sheet organized structure. The random coil structure can easily change into a β-sheet structure because of moisture absorption. Water cannot prolong moisturization of skin, because of its short residence in the skin and on its surface. Sericin enhances the water-holding capacity of the skin possibly attributed to the restoration of amino acids.[20]

In addition to making up a protective film on the skin, whole globular silk protein also produces a feeling of elasticity and keeps the skin hydrated properly. Silkall™ 100 (Ikeda Corp.) is a purified grade of natural silk, processed to retain its original physical structure and chemical composition. The processed silk powder can retain and release moisture depending on the ambient environment humidity and temperature. Fibro-Silk™ is also a processed silk protein that can entrap oil and is used in shampoo to give shiny hair.

2.3.3.4 *Milk Proteins*

Basic milk proteins are casein, the most abundant, constituting up to 80% of the proteins in milk, lactoglobulins and lactoalbumins. Lactoglobulins are unique due to their good solubility in water. Milk is one of the oldest and most famous naturally occurring skin softeners and its effectiveness and benefits in the world of beauty and personal care have long been known.[23] Milk proteins have been used in aqueous and aqueous–alcoholic hair care formulations, with the purpose of providing preventive care of scalp and hair. Follicusan™ is a bioactive signaling milk protein. It has been demonstrated to vitalize scalp cells, prevent formation of dandruff by reducing excessive sebum secretion and counteract premature hair loss.[24] Milk proteins can also reinforce the moisturization of the skin and provide suppleness and comfort. Lactokine™ minimizes skin irritation and hastens the recovery of skin by preventing the release of inflammatory mediators.

2.4 Production of Hydrolyzed Proteins

2.4.1 Major Steps

Proteins that are extracted from different origins have different properties. For hydrolyzed proteins however, their methods of production and preparation also play an important role in determining their properties. Their production and processing are generally held as business secrets that are not available to the public. However, the major steps involved can be described. The first step is the extraction of proteins. Normally proteins do not exist in their pure form, as they always coexist with some other substances in organisms. Protein extraction is therefore the first process in the preparation of protein-based raw materials for cosmetics. The protein extraction method for plant proteins is similar to that used in the food industry. Extraction methods that can be applied to food industry waste such as blood plasma, meat waste and osseous ore have also been developed. Some proteins that exist in pure form can be found in specific organs and tissues and require no purification. Proteins that coexist with carbohydrates such as starch are technically difficult to extract. The final product from starch consists of both hydrolyzed proteins and carbohydrates. After the proteins are extracted, they will undergo hydrolysis. However, some peptide bonds are not available to hydrolyzing factors due to the 3D structure of the protein. In such cases, a chemical or thermal treatment to denature the proteins is performed. After hydrolysis, the products are purified by adsorption and further concentrated in vacuum. Conditioning and the addition of preservatives is the last step, which could take days or weeks. The final product will be liquid protein hydrolysates with high water solubility and low organic solvent solubility with wide range of molecular weights.

2.4.2 Hydrolysis of Proteins

Hydrolysis is the breaking of peptide bonds. Hydrolysis of proteins can be carried out using acids, bases or enzymes. The reaction depends on the amount of hydrolyzing agent, temperature and duration. In enzymatic hydrolysis, the type of enzyme and its activity are the determining factor in the reaction.

2.4.2.1 Acid Hydrolysis

Amino acids may be lost during chemical hydrolysis. For example, asparagine and glutamine can be converted into aspartic and glutamic acid under acidic conditions, reducing the isoelectric point of the product. Tryptophan degrades into smaller molecules completely and tyrosine only undergoes partial hydrolysis, with loss of serine and threonine. Bonds between aliphatic amino acids are relatively stable against chain scission, such as lle–lle, Val–Val, or Val–lle. Hydrochloric acid is generally applied in acidic hydrolysis for ease of separation. Sulfuric acid, which is strong and non-oxidizing, is also used. Methane-sulfonate, *p*-toluenesulfonate acid and a mixture of them may also be applied.

2.4.2.2 Base Hydrolysis

Base hydrolysis is less used than acid hydrolysis. Base hydrolysis does not chemically modify tryptophan, however serine is not stable under basic conditions. Serine may be converted to alanine and glycine. Deamination of arginine produces ornithine and citrulline. Cysteine may decompose completely with production of hydrogen sulfide and ammonia. The hydroxide of sodium, calcium, potassium and barium can all be used in base hydrolysis. It is most frequently applied in protein mixtures with a high content of carbohydrates. Protein solutions for pharmaceutical applications are also produced by base hydrolysis.

2.4.2.3 Enzymatic Hydrolysis

Enzymatic hydrolysis methods started to replace previous chemical hydrolysis processes and are now the most widely used processes. Enzyme hydrolysis can take place under much milder conditions than chemical hydrolysis. Minimal heating is needed and pH values can be moderate and should be similar to physiological pH conditions. Some enzymes have the highest activity in alkaline solution, such as chymotrypsin (pH = 8). Some enzymes have the highest activity in neutral pH, such as bromelain. Some enzymes have the highest activity in acidic conditions, such as pepsin (pH = 2). Enzymatic hydrolysis can be controlled by adjusting pH values or temperature. Most of the enzymes used are proteases obtained from animal and plant sources, such as papain, chymotrypsin and pepsin. Microbes and fungi can also serve as enzymes. Some alkaline enzymes can be obtained from *Bacillus*. A mixture of proteases or proteases with amylases can be used in hydrolysis, especially in acidic conditions.

2.4.2.3.1 Enzymes Used. Trypsin can break peptide bonds that are situated at the carboxylic group of lysine or arginine. When there is access to the functional groups of lysine or arginine, the hydrolysis becomes selective. Thrombin is more reactant specific than trypsin. The peptide bonds between glycine and arginine are specifically reactive with thrombin. Chymotrypsin disrupts peptide bonds with carboxylic groups on aromatic amino acids. For long peptide chains, chymotrypsin attacks peptide bonds among methionine, asparagine, valine and leucine. Pepsin is not reactant specific; it breaks peptide bonds between tryptophan, leucine, methionine and phenylalanine. Subtilisin breaks bonds that are adjacent to serine, glycine and aromatic amino acids. Elastin breaks peptide bonds between neutral amino acids. Papain breaks peptide bonds among arginine, lysine and glycine. Peptide bonds between acidic amino acids are stable against papain. Thermolysin breaks peptide bonds between amino acids that have a hydrophobic side-chain.

2.5 Discussions and Comparisons

2.5.1 Molecular Weight of Proteins

Molecular weight is an important parameter determining the properties of the proteins for cosmetic products. Generally, the average molecular weight of hydrolyzed proteins can be calculated by size exclusion chromatography, which is also known as gel permeation chromatography. The method is based on the fraction of peptides in terms of molecular weight. The size and length of peptides depends on the level of disintegration in hydrolysis of the proteins. The level of disintegration is represented by the hydrolysis level, which is written as $\mathrm{DH} = h/h_{\mathrm{total}}$, where h is the number of peptide bonds broken per unit mass of the protein and h_{total} is the total number of peptide bonds per unit mass of the protein. Apart from size exclusion chromatography, the most commonly used method to determine DH is by measuring the nitrogen produced during hydrolysis in the presence of trichloroacetic acid. The quantity is determined by titration using ninhydrin, fluoresceinamine or trinitrobenzene sulfonic acid. A list of molecular weight of proteins can be found in Table 2.5.

2.5.2 Isoelectric Point

Proteins and protein hydrolysates have ampholytic properties because they contain both carboxylic groups and amino groups at free end positions and functional groups. These groups can be ionized. The charge of the proteins is influenced by the 3D structural arrangement and the distribution of these carboxylic groups and amino terminal positions. Based on the pH value of the environment, the protein molecules exhibit a positive charge in acidic conditions and a negative charge in basic conditions. The isoelectric point is when the overall charge of the proteins is zero due to balanced formation of zwitterions. Therefore, the isoelectric point is a measure of the relative content of acidic and basic amino acids and their dissociation constant. If a protein is made of more basic amino acids, the isoelectric point will be in the alkaline range of pH, and if most of the amino acids in the protein are acidic, the isoelectric point will be in the acidic range of pH. The isoelectric point value of some proteins is characteristic and therefore can be used to estimate the purity of the proteins. The isoelectric point of selective proteins can be found in Table 2.6.

After hydrolysis, generally the isoelectric point will be changed as compared to the original protein. Hydrolysates of the same protein may have different isoelectric values depending on the peptide chain length and hydrolysis conditions. Table 2.7 gives a summary of the isoelectric point of the hydrolysates of some proteins with different molecular weights. The ionic nature of many proteins and their hydrolysis products will determine the interactions with the surfaces of skin or hairs. Therefore, the isoelectric point and the pH value of the cosmetic product are important in determining its effectiveness. The determination of the isoelectric point can be done by plotting a titration curve which reflects the ionization state of the molecules,

Table 2.5 Molecular weights of selected proteins.

	Protein	Molecular weight [Da]
Collagen	Tropocollagen	28 5000
Keratin	α-Keratin	50 000
Elastin	Tropoelastin	72 000–74 000
Silk proteins	Fibroin	370 000, 25 000
	Sericin	65 000–400 000
Egg protein	Ovalbumin	44 500
Wheat gluten	Gliadin	25 000–150 000
	Glutenin	40 000–100 000
Soy proteins	Glycinin, conglycinin	20 000–400 000
Milk proteins	Casein	19 000–24 000
	Lactoalbumin	14 200
	Lactoglobulin	18 300

Table 2.6 The isoelectric points for selected proteins.

Protein	Isoelectric point
Collagen	7.0–7.8; 6.7
Keratin (hair cuticle)	3.7
Milk lactoalbumin	5.1
Milk lactoglobulin	5.3
Milk casein	4.6
Egg ovalbumin	4.5
Silk fibroin	5.1

Table 2.7 The isolectric point (pI) value of selected protein hydrolysates.

Protein	Molecular weight [Da]	Isoelectric point	Protein	Molecular weight [Da]	Isoelectric point
Collagen	64 000	7.3	Keratin	7800	4.8
Collagen	22 000	5.6	Keratin	7500	5.2
Collagen	1850	7.1	Wheat gluten	9500	5.3
Collagen	1700	6.6	Wheat gluten	5600	5.2
Collagen	1650	5.4	Wheat gluten	3700	5.6
Elastin	31 000	6.6	Wheat gluten	3500	3.9
Elastin	12 500	6.3	Wheat gluten	2200	4.9
Elastin	9000	5.8	Wheat gluten	1800	4.7
Elastin	8500	5.4	Corn gluten	1700	5.2
Keratin	46 500	5.2	Corn gluten	800	4.6

or by using electrophoresis in different pH conditions or by electrofocusing. Ionic chromatography is used to determine the average isoelectric point.

2.5.3 Hydrophobicity

Protein structure formation is greatly affected by non-covalent hydrophobic interactions. Hydrophobicity also affects the substantivity, thin film forming and reducing irritation potential of surfactants. Hydrophobicity can be determined experimentally, however generally theoretical calculation of hydrophobicity of different proteins is good enough to give an indication. The content of a particular amino acid hydrophobicity, its position (if it is at a terminal position) and the peptide chain length should be taken into consideration. For protein hydrolysis products, a particular amino acid's position in the main chain plays a more important role on hydrophobicity than in the case of high molecular weight complex proteins. Short chain peptides with hydrophobic amino acids in terminal positions are more hydrophobic than those peptides in which hydrophobic amino acids are in the central part of the chain. In the case of low molecular weight peptides with hydrophobic amino acids at terminal positions, it is not possible to arrange the polar section of the chain in a manner such that it is not accessible to water. However, for long chains, the arrangement of the long chain will protect polar amino acids from contact with water.

Hydrophobic interactions influence the formation of peptide aggregates. The aggregates of polymer chains decrease water solubility, especially in the case of hydrolysates from proteins with a high concentration of non-polar amino acids. The hydrophobicity of hydrolysates may be controlled by the selection of enzymes used to target peptide bonds between certain amino acids.

2.5.4 Comparison of Different Hydrolysis Processes

The selection of hydrolysis methods is mainly based on production costs. The requirements of molecular weight and amino acid content, especially at terminal positions, are also taken into consideration. Chemical methods are less selective than enzymatic methods because all the peptide bonds have equal chance of contact with H^+ or OH^- ions. The amino acids at the terminal positions have a much better chance of contact with the hydrolyzing agent. Therefore, hydrolysates obtained from chemical hydrolysis contain free amino acids and this method is employed to produce low molecular weight products. Enzymatic methods can only produce products with molecular weights larger than 500 Da. Amino acids at the terminal position can be tailored by the selection of enzymes used because specific enzymes target specific peptide bonds between specific amino acids. The molecular weight can also be tailored by the enzyme specificity, which is not easily attained by chemical synthesis. The peptides obtained by enzymatic hydrolysis have molecular weights ranging from 500 to 10 000 Da, in contrast to 500 to 30 000 Da in peptides obtained from chemical hydrolysis. The color and odor of enzymatic hydrolysis products are better in comparison with products produced by the chemical process.

2.6 Outlook and Prospects

Consumers' trend towards healthier lifestyles has been the driving factor for the use of natural polymers in personal care products. Good health is often related with products produced by nature as opposed to those produced through artificial chemical means. This perception has been reinforced by negative publicity about some synthetic chemicals used in cosmetics. The verified or unverified press coverage on some of the chemicals' carcinogenic effects also had a very important impact on market demand, driving consumers to natural alternatives. Consumers are attracted to cosmetics containing natural additives and ingredients not just because of their assumed and perceived benefits on health; many people believe that ingredients extracted from natural products are also much better quality than synthetic chemicals. This perception enables cosmetic manufacturers to charge premium prices for natural cosmetics with better packaging that highlight their "natural" credentials. This has enabled natural personal care products to expand quickly into upscale establishments, and there has been a growth of "natural supermarkets" that specialize in the sales of natural personal care products with a wide range of products, including toothpaste. The market demand for natural personal care products will increase but overall sales will remain weaker than that of mainstream products. The trend will be to increase the content of natural ingredients to satisfy consumers' expectation. However, synthetic chemicals will not be completely abolished. Natural additives and ingredients used will be further highlighted by the major manufacturers. Natural ingredients and major technologies by the personal care industry largely remain confidential. Many synthetic ingredients are found to be safer than natural ingredients because of a long history of usage. As the market further develops, regulatory authorities will push manufacturers to disclose ingredients and processes used as well as safety evaluations with verifiable information.

References

1. U. S. F. a. D. Administration, Cosmetics Safety Q&A, *Personal Care Products*, http://www.fda.gov/cosmetics/resourcesforyou/consumers/ucm136560.htm, accessed 10 Oct 2015, 2015.
2. S. I. T. S. Portal, *Breakdown of the cosmetic market worldwide from 2011 to 2014, by product category*, http://www.statista.com/statistics/243967/breakdown-of-the-cosmetic-market-worldwide-by-product-category/, accessed 10 Oct 2015, 2015.
3. M. S. J. A. K. Pytkowska, *SOFW J. Wydanie Polskie*, 2008, **1**, 12–16.
4. Y. Komaki, *Fragrance J.*, 1994, **22**, 80–88.
5. R. C. Potter, J. M. Castro and L. C. Moffatt, US5620692, 1997.
6. Y. Chen and Y. Hua, *Riyong Huaxue Gongye*, 2000, **30**, 62–64.
7. R. Voegeli; K. Stocker and C. Mueller, US5322839, 1994.
8. W. Wu and N. S. Hettiarachchy, *J. Surfactants Deterg.*, 1998, **1**, 241–246.
9. K. Nosaka, *Fragrance J.*, 1998, **26**, 19–24.
10. Z. Nawaz, WO1998055089A1, 1998.
11. M. Yoshioka and Y. Kamimura, Japan Pat., JP 04139115, 1992.
12. M. E. Chvapli and Z. Eckmayer, *Int. J. Cosmet. Sci.*, 1985, **7**, 41–49.
13. A. Huc, D. Hartmann, M. Journot, P. Comte and G. Ville, *Int. J. Cosmet. Sci.*, 1981, **3**, 159–183.
14. T. Nakane, *Fragrance J.*, 1999, **27**, 58–68.
15. B. J. Gesslein and R. T. Smith, *Cosmet. Toilet*, 1987, **102**(**52**), 56.
16. M. Yoshioka; H. Shintani; M. Ai-e and A. Kei, Japan Pat., JP 2003040727, 2003.
17. M. Philippe, J. C. Garson and J. P. Arraudeau, US6841162, 2005.
18. N. G. Ikeda, N. Goto and M. Yoshioko, *Fragrance J.*, 2003, **31**, 113–118.
19. S. Daikai, S. Nagao and S. Uchida, *Fragrance J.*, 2000, **28**, 65–71.
20. M. N. P. Padamwar, P. Atmaram, V. Daithankar Aarti and K. R. Mahadik, *J. Cosmet. Dermatol.*, 2005, **4**, 250–257.
21. R. M. Patel and M. K. Modasiya, *Int. J. Res. Pharm. Biomed. Sci.*, 2011, **2**, 913–917.
22. S. H. A. Pomerantz and I. G. Ances, *J. Clin. Invest.*, 1975, **55**, 1127–1131.
23. Y. F. Tamura and Y. Fukuwatari, *Fragrance J.*, 1991, **19**, 119–121.
24. H. Negishi, H. Otomo, T. Gotou, T. Ueda and T. Kuwata, *Cosmetic properties of whey minerals and their application to skin care products for babies*, International Dairy Federation, 1998, pp. 337–350.

CHAPTER 3

Polyacrylates for Personal Care

IVAN WIYANTO[b] AND XIAN JUN LOH*[a,b]

[a]Institute of Materials Research and Engineering, A*STAR (Agency for Science, Technology and Research), 2 Fusionopolis Way, Innovis, #08-03, Singapore 138634, Singapore; [b]Department of Materials Science and Engineering, National University of Singapore, Singapore 117574, Singapore
*E-mail: lohxj@imre.a-star.edu.sg

3.1 Introduction

Personal care products, also more widely known as cosmetics, are defined as products which not only can improve the look of an individual's skin, hair, nails or teeth, but also can enhance the self-esteem of someone under medical treatment by concealing the flaws. Recently, the demand for personal care products is burgeoning as people are becoming more aware of staying hygienic.[1,2]

In the second half of the twentieth century, polymers have been used extensively in personal care products. The polymers used in personal care products range from both synthetic and natural polymers, although there is a recent paradigm shift that using synthetic polymers is less favourable due to their non-environmentally friendly solvents. This problem can however be resolved by the use of aqueous solvents like water.[3] The high demand to use polymers in personal care products is mainly because it is easy to obtain various final performance product attributes just by altering the structures and properties of the polymers (*i.e.* by cross-linking, copolymerisation and branching).[1]

RSC Polymer Chemistry Series No. 20
Polymers for Personal Care Products and Cosmetics
Edited by Xian Jun Loh

Published by the Royal Society of Chemistry, www.rsc.org

Polymers in personal care products play important roles as thickeners, rheological modifiers, media for delivery of active ingredients, film formers, emulsifiers, moisturisers, dispersants, waterproofers, conditioners and the list goes on.[1,4] Their key functionalities include the ability to increase the viscosity of solutions, form physical gels, stabilise dispersions and emulsions, induce particle aggregation, solubilise hydrophobic compounds, modify surface properties and facilitate the controlled release of active compounds.[5]

One of the polymers used in personal care products is *polyacrylates*. The term "acrylates" is defined as non-cross-linked copolymers containing a combination of acrylic acid, methacrylic acid and their simple esters.[4,6,7] Figure 3.1 shows some examples of monomers based on acids. One common feature of the monomers is the presence of the vinyl (C=C) bond. Polyacrylates are mainly used as rheological modifiers in cosmetics, in which they are added to thicken and control the viscosity in the formulation process.[1,8,9] Polyacrylates are readily available worldwide from many suppliers like BASF and have a competitive advantage in terms of their price relative to some other alternatives. The market size for polyacrylates is also quite large. In 2005, the market for polyacrylates reached around 3.2 million tons and 2 million tons were contributed by the cosmetic products. The figure is predicted to keep growing in the future.[10,11] Furthermore, polyacrylates can be produced in various physical forms (solid, oil-based liquid and oil-in-water emulsions) with different ionic charge types and densities to cater to the requirements of various uses.

Rheology is a field of study investigating deformation and flow of matter. The rheology of a polymer depends on the extent of the force applied to it and also the intramolecular forces that hold the polymers together.[8,12] In cosmetics, rheological modifiers function to prevent sedimentation of the dispersed pigments in the formulation during storage and also to provide desirable delivery characteristics upon application to human body so as to add to their aesthetic values and meet the sensory expectations of the customers.[9,12,13]

As seen from Figure 3.2, rheological modifiers can be classified into two big groups: inorganic and organic.[13] *Inorganic rheological modifiers* are made of inorganic products like clays or silicas and typically used as secondary thickeners. On the contrary, *organic rheological modifiers* are classified further

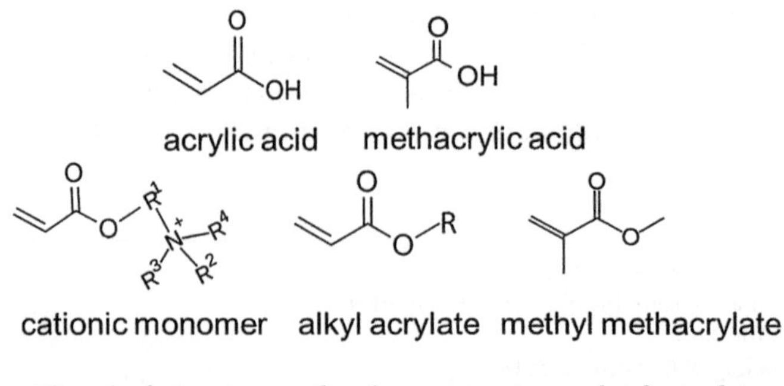

Figure 3.1 Chemical structures of various monomers of polyacrylates.

into the natural and synthetic ones. Previously, natural polymers such as cellulose and xanthan gums were used to modify the viscosity of cosmetic products. As technological advancements arose, synthetic polymers like polyacrylates and polyurethanes have been used more widely as rheological modifiers as their architecture can be easily modified through various chemical reactions. In light of environment concerns, industries today produce novel rheological modifiers from natural and renewable resources.[8]

Synthetic organic rheological modifiers can be further divided into two main groups: associative and non-associative. *Associative rheological modifiers* thicken polymers by non-specific interactions of hydrophobic end-groups with themselves and also with the components of the cosmetics by forming a large physical network. This network causes the viscosity of the polymer to increase. Some examples of associative rheological modifiers are polyurethanes and polyethers. Meanwhile, *non-associative rheological modifiers* form entanglements of water-soluble, high molecular weight polymer chains, also known as hydrodynamic thickening. Polyacrylates, the highlight of this chapter, fall under both associative and non-associative rheological modifiers.[13]

It is important for cosmetic products to have their rheological properties modified, mainly in order to suspend the dispersion of additives in the fluid and to improve the dispersion stability. It is also important in assessing if the flow properties of the product will be meaningful when produced, ensuring consistency, smoothness and spreadability. Moreover, not only the physical stability of the products will be affected by the rheological properties, but also their biological stability.[14] As such, the deep understanding of rheological concepts is vital in the cosmetic industry.

It should be noted that different liquid polymer systems have different abilities to resist the force applied to them. The resistance of the fluid to flow is usually referred to as *viscosity*. Upon application of the force, the system will flow to relieve the strain due to the force. The rheological characteristics of a liquid can be classified into two big groups: Newtonian and non-Newtonian.[13,14] Systems exhibiting *Newtonian* properties have constant viscosity when stressed. Consequently, the flow velocity is directly proportional to the shear stress. Examples of systems with Newtonian properties are pure water, solvents and mineral oils. On another note, most cosmetic products display *non-Newtonian* rheological characteristics, whereby the

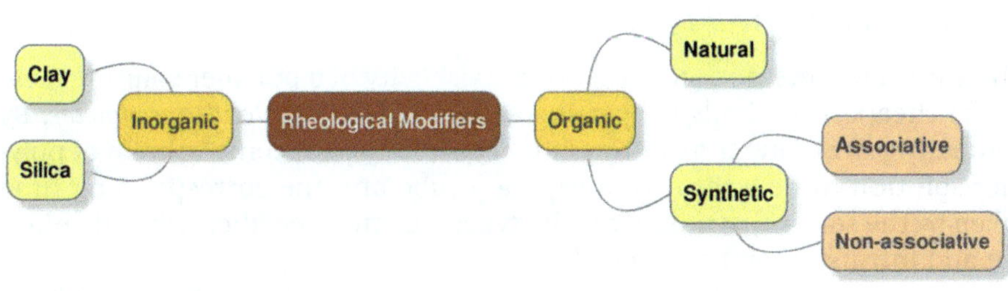

Figure 3.2 Classification of rheological modifiers.

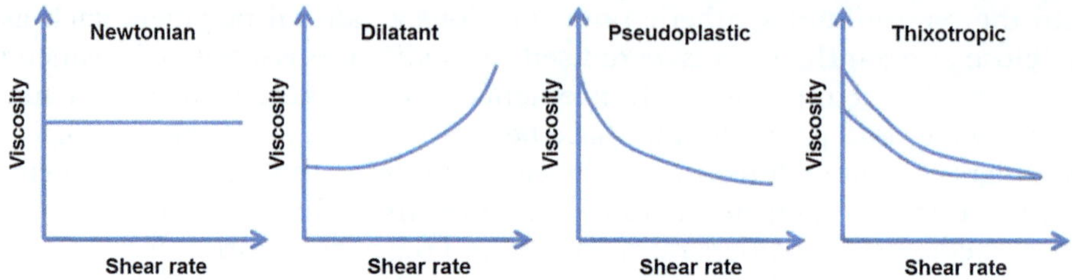

Figure 3.3 Viscosity–shear rate curves of different rheological profiles.[13]

viscosity is dependent on the shear stress. There are three subcategories of non-Newtonian characteristics: dilatant (shear thickening), pseudoplastic (shear thinning) and thixotropic (time dependent rebuilding of viscosity).[11,15] Figure 3.3 summarises the different rheological profiles in terms of their viscosity *vs.* shear rate plots.

The rheological properties of many polyacrylate polymers are influenced by some factors. The most prominent factor is the presence of a viscosity modifying agent. For example, ions can reduce the viscosity of polyacrylate polymers. Other factors would be concentration of acid groups, molecular weight, cross-linking degree, solids content, pH value, viscosity of internal and external phases, volume ratio of the phases, particle size distribution, type and amount of emulsifiers used during synthesis. The ratio of the ingredients added also can affect the rheological characteristics of the solvent dramatically.[13,16]

3.2 Uses of Polyacrylates

Polyacrylate-based polymers have many uses in cosmetic products. When polyacrylate is used as film former, binder and fixatives, the concentration of polyacrylate in a formulation can be higher than 25%. It can be as low as 0.5% if it is used as suspending agent, emulsion stabiliser or aqueous viscosity-increasing agent.[6] The different uses of some polyacrylate-based polymers are summarised in Table 3.1.

3.2.1 Characterising Rheological Modifiers

3.2.1.1 Rheological Effects

There are a few methods to measure the viscosity of a polymer solution. One of the cheapest methods is by using a glass *U-tube Ubbelohde viscometer*. By comparing the efflux time or the time required for a specific volume of polymer solution to flow through a capillary tube and the corresponding time taken by the solvent, various viscosity types (relative, specific, reduced, inherent and intrinsic) can be defined.[8,17,18]

There are a few other conventional types of viscometers, which have been used to measure viscosity. In the past, the easiest method was to use a *Ford*

Table 3.1 Different uses of selected polyacrylate-based polymers[7,70]

Polymer	Function	Common product category
Acrylates copolymer	Binder, hair fixatives, film former, nonsurfactant suspending agent, aqueous viscosity-increasing agent	Mascara, hair dyes and colours, blushers, face powders, lipstick, base-coats, undercoats, manicuring preparations
Ammonium acrylates copolymer	Binder, aqueous viscosity-increasing agent, film former	Eyeliner, mascara
Sodium acrylates copolymer	Binder, aqueous viscosity-increasing agent, film former	Hair dyes and colours
Acrylates/PVP copolymer	Binder, hair fixatives, film former, nonsurfactant suspending agent	Tonics, dressings, hair grooming aids, wave sets
Poly(acrylic acid)	Binder, aqueous viscosity-increasing agent, film former, emulsion stabiliser	Nail polish, enamel, bath soap, detergent, cleanser, body and hand preparations, plate masks
Ammonium polyacrylate	Emulsion stabiliser, film former	Eyeliner
Potassium aluminium polyacrylate	Absorbent, aqueous viscosity-increasing agent, binder	Absorbent
Sodium polyacrylate	Film former, aqueous viscosity-increasing agent, hair fixative	Hair spray, shampoos, bath soap, detergents, skin care products
Styrene/acrylates copolymer	Film former	Hair dyes and colours, eyeliner, permanent waves, nail polish, enamel, face and neck preparations
Ethylene/methacrylate copolymer	Film former	Blushers, foundations, make up bases

flow cup in which the viscosity is calculated based on the time taken for a fixed volume of fluid to flow through a tube having fixed dimensions. However, the temperature can become uncontrolled.[19] Then, the *falling ball viscometer* was discovered to overcome this problem. In this method, a ball is dropped in a tube and the motion of the sphere can be detected by either an ultrasound method or laser light through a window. Once again the validity of the result is limited by the roughness and elasticity of the sphere used, which caused a 26% error in the viscosity measurement.[20]

Thanks to advancements in the technology, there are now precalibrated viscometers that record values at various rotational speeds, like the *Brookfield viscometer*. The surface of these instruments rotates about a common axis while the other axis stays stationary. It measures the torque required to rotate the spindle in a fluid. The viscous drag is proportional to the speed of the spindle's rotation.[13,21] The current version of this viscometer can even measure viscosity at precise time intervals set by the experimenter. However, one limitation regarding this method is that it only covers the low to medium shear range. Similarly, other viscometer methods such as the *Krebs Stormer viscometer* only measure rheology at the medium shear range and the *ICI cone and plate viscometer* covers only the high shear range.[13] A viscometer is only useful when the test only requires simple flow measurements.

Sometimes, in order to understand the complete rheological profile, it is desirable to take measurements from the low, medium and high shear ranges. A *rheometer* can be used for this purpose, as it allows greater characterisation of flow and deformation behaviour. An advanced rheometer is even able to measure various ranges of shear viscosities under different modes (*e.g.* constant shear stress, constant shear rate, *etc.*) precisely.[13] A rheometer is much more expensive than viscometers, and thus it should be used only when appropriate.

3.2.1.2 Other Characterisations

As mentioned previously, there are some other parameters that are of particular interest in studying the rheological properties of cosmetic polymers. The particle size distribution can be measured by the *dynamic light scattering* technique, where the sample is illuminated using a laser beam and a fast photon detector detects the fluctuation of the scattered light at a known scattering angle.[22] Alternatively, one can investigate the particle size distribution by *capillary hydrodynamic fractionation*. In this technique, colloidal particles of differing sizes are dispersed and carried through an open capillary in a fluid. Fractionation causes them to emerge in order of decreasing diameter. This technique shows a good agreement within 5% with transmission electron microscopy (TEM) in particle size diameter measurement.[23]

The molecular weight and molecular weight distribution of the polymers can be measured by *gel permeation chromatography* (*GPC*). GPC establishes a quantitative relationship between a polymer's molecular weight and the elution volume.[24] The polymer is first dissolved in an appropriate solvent (*e.g.* tetrahydrofuran) and subsequently the aqueous polymer is passed though a stationary phase column containing swollen gel that acts as a molecular sieve. Polymer molecules having a low molecular weight can penetrate into the gel particle pores. Meanwhile, the high molecular weight polymer molecules are excluded from the pores and elute first. The elution time based on a calibrated standard curve determines the molecular weight and molecular weight distribution of the polymer.[25]

To measure the *solids content* of a polymer emulsion (nearly equal polymer content), 1 gram of the sample can be weighed in a flat-bottomed glass dish. The sample is tilted gently and spread for about 3 hours in a ventilated oven at a suitable elevated temperature. The volatile matter will be evaporated and the solid content remaining determines the percentage of solid content contained in the emulsion by following formula (3.1) below:[26]

$$\text{Solids content (\%)} = \frac{(C - A)}{B} \times 100 \tag{3.1}$$

where A = weight of empty dish (g), B = weight of sample used (g) and C = weight of dish and sample after heating (g).

3.3 Monomers of Polyacrylates and Their Synthesis

As mentioned previously, one common monomer of polyacrylates is *acrylic acid*. The synthesis route of acrylic acid starts from propylene, the primary feedstock derived from crude oil. The unique feature of polypropylene is the coexistence of a double bond for polymerisation and the dangling functionalisable $-CH_3$ bond.[8]

Firstly, propylene is converted into acrylonitrile *via* catalytic vapour-phase ammoxidation.[27] The reaction scheme is as follows: $CH_2{=}CHCH_3 + NH_3 + 3/2\ O_2 \rightarrow CH_2{=}CHCN + 3H_2O$. Afterwards, acrylonitrile is reacted with a small excess of sulfuric acid $H_2SO_4{\cdot}H_2O$ in glass-lined equipment at 90–100 °C for 1 hour. The reaction scheme is: $CH_2{=}CHCN + H_2SO_4{\cdot}H_2O \rightarrow CH_2{=}CHCONH_2{\cdot}H_2SO_4$. The by-product separation from its sulfate salt can be performed by a base neutralisation method or an ion exclusion column. This recovery process is usually the most challenging and the most costly part of the synthesis. When a solid crystalline monomer is wanted, the base neutralisation method with ammonia should be chosen to yield ammonium sulfate. The ion exclusion method utilises a sulfonic acid ion-exchange resin. A dilute solution of acrylamide would be produced in water. However, it should be noted that the main concern in the recovery step is the waste production which might cause pollution.[28] Lastly, the reaction mixture will be purified using approximately 10 theoretical stages in a tower or centrifugal extractor[29] and aqueous acrylic acid can be recovered. The complete synthesis route of acrylic acid is represented by Figure 3.4.

The above synthesis route to produce acrylic acid is the most conventional way adopted by many companies, even though it is not the most efficient pathway. In order to be commercially attractive, the raw material and utilisation costs must be low and waste disposal charges must be minimal.[27] However, there are some drawbacks of the method which make it somehow less attractive for large scale production: a very high temperature is needed, the presence of side reactions, a risk of radical-initiated exothermic polymerisation with the likelihood of acrylic acid dimerisation and environmental concern about the waste products.[30] As such, a greener method was developed by Kamal *et al.* to produce acrylic acid *via* direct bioconversion of acrylonitrile by *Rhodococcus ruber* strain AKSH-84 in a medium having pH 7.0 at 30 °C. The

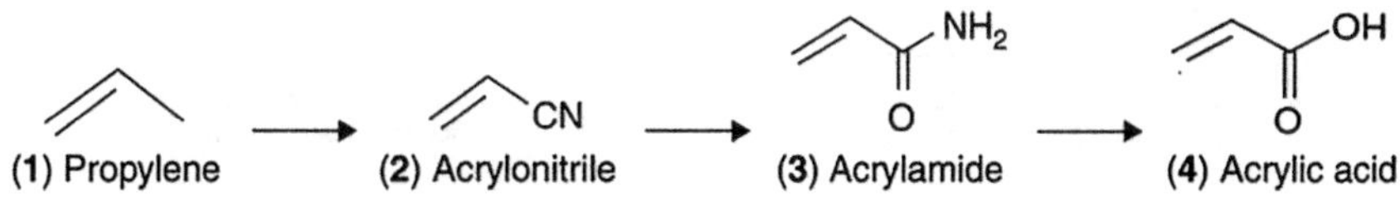

Figure 3.4 Common synthesis route of acrylic acid from propylene.[8]

yield of the bioconversion using purified nitrilase (50 U mg^{-1}) as the enzyme which catalyses the reaction shows a better result of 92% compared to using whole resting cells which only has a yield of about 63%.[11] The comparison of the result is clearly depicted in Figure 3.5.

Polymerisation of acrylic monomer is prone to going uncontrolled in the presence of species that cause 'initiation,' even in small amounts. This can be circumvented by addition of a stabiliser, which is commonly referred to as an 'inhibitor.' One common inhibitor is paramethoxyphenol (PMP), which works most efficiently when a trace amount of dissolved oxygen is present in the water. When polymerisation is desired, the dissolved oxygen may be purged by introducing an inert gas such as nitrogen. As a consequence, the inhibitor will work less effectively and a controlled polymerisation starts to happen.[8]

Acrylic acid can be further functionalised to form *alkyl acrylate* by reacting it with alcohol *via* acid-catalysed esterification. The general equation for this reaction is: acrylic acid + alcohol ⇌ alkyl acrylate. *Monomers having cationic groups* (*i.e.* a quarternary aluminium salt) can also be produced by trans-esterification reaction of short chain alkyl acrylate with an alcohol having a diamine group followed by further quarternisation with a quarternising agent.[8]

Another common monomer of polyacrylates, *methacrylic acid*, can be synthesised by hydrolysing methacrylamide sulfate, which is obtained from acetone cyanohydrin.[31,32] Acetone cyanohydrin is produced by the reaction of dry acetone and hydrogen cyanide with basic catalyst. The final reaction product is reacted at 80–110 °C with excess concentrated sulfuric acid solvent of about 1.5 molar equivalent to cyanohydrin to yield methacrylamide acid sulfate. Finally, the product is cracked at 150 °C for 1 hour to obtain the methacrylamide sulfate. The next step would be hydrolysis of methacrylamide sulfate with excess water to form methacrylic acid. As a by-product, ammonium acid sulfate will be formed. Alternatively, methanol can be used to replace water in a hydrolysis–esterification step which yields a mixture of methacrylic acid and *methyl methacrylate*, another monomer for polyacrylates.[33]

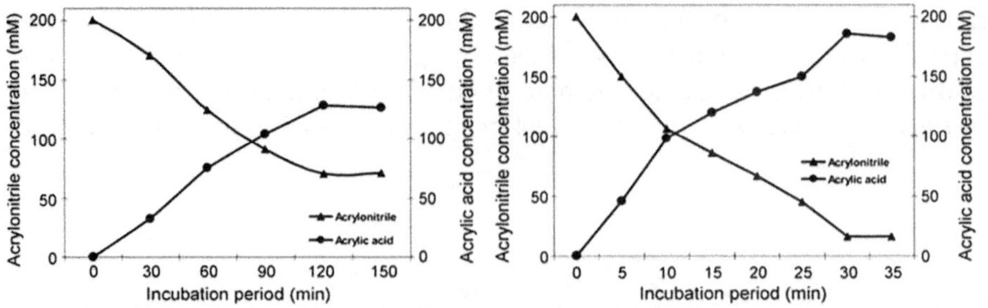

Figure 3.5 Bioconversion profile of acrylonitrile to acrylic acid using purified whole resting cells (left) and using purified nitrilase (right).[11]

3.4 Alkali Swellable Emulsion (ASE)

One example of acrylic-based non-associative thickeners is alkali swellable emulsions (ASE). The mechanism of hydrodynamic thickening by ASE is as follows: a long chain acrylic polymer, for example a copolymer of methacrylic acid and ethyl acrylate[34] (Figure 3.6), is supplied at low pH and thus exists as a dispersion of coiled molecules. Upon neutralisation (*i.e.* when the pH is raised to >7), the pendant carboxylic groups in the thickener start to dissociate. Consequently, the polymer 'swells,' becomes more water soluble and begins to uncoil. A further increase in pH causes the polymer to adopt a more open long-chain structure and the molecules get entangled with each other, raising its viscosity.[13,15] This reaction mechanism is depicted in Figure 3.7.

The most common ASE is called poly(methacrylic acid-ethyl acrylate) or P(MAA-EA) for short. P(MAA-EA) is synthesised through semicontinuous emulsion polymerisation of MAA and EA monomer with a weight ratio of 1 : 3, as well as a diallyl phthalate cross-linker in a small amount (0.3 mol% of the entire monomers). Additionally, a pre-emulsion containing monomers,

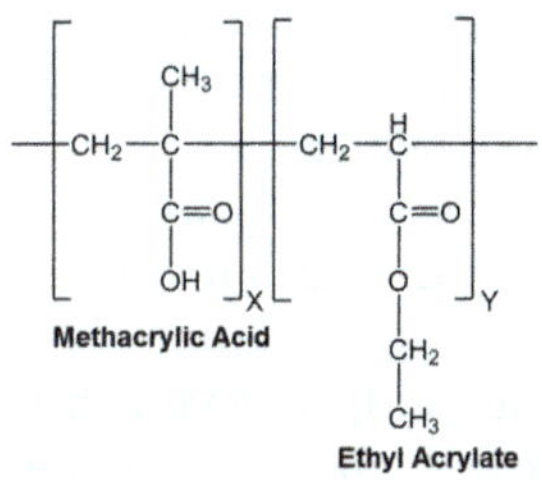

Figure 3.6 Structure of a common ASE, P(MAA-EA).[34]

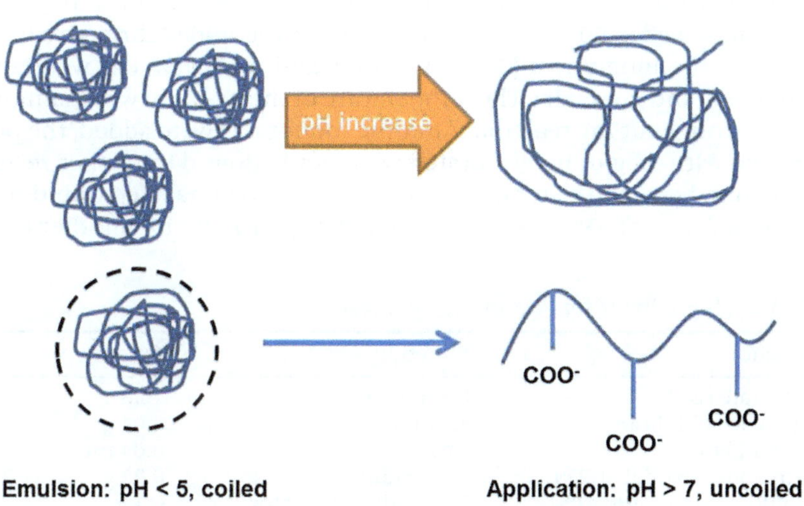

Figure 3.7 Mechanism of hydrodynamic thickening in an ASE.[13]

sodium dodecylsulfate (SDS) which acts as the surfactant and ammonium persulfate (APS) which acts as the free radical initiator was prepared. The reaction begins when 5 wt% of the pre-polymer is added to the flask and then it is allowed to proceed at 86 °C for 45 minutes. Soon afterwards, the remaining pre-emulsion is added and the reaction continues for another 1 hour before the temperature is increased to 92 °C for a further reaction of 20 minutes. Finally, the reaction is cooled down to room temperature. The full ingredients for the synthesis of P(MAA-EA) are presented in Table 3.2.[35]

The synthesis of ASE itself possesses a few drawbacks, mostly related to the use of surfactants. Surfactants have the most important function of stabilising the polymer particles by forming micelles while suspended in water so as to improve the polymerisation rate.[36] It is critical to choose the best surfactant to synthesise a polymer using emulsion polymerisation. However, it is difficult to remove surfactants from the final product. The use of surfactants also unavoidably causes these difficult-to-remove foams to form, which might degrade not only the appearance of the final product, but also the efficiency of the thickener as the physical properties of the resulting polymer will be adversely affected. Even when the surfactants can be removed by a desorption method, coagulation or flocculation of the polymer might happen, which is undesirable.[37] Moreover, the high cost of surfactant discourages its excessive usage in ASE synthesis.

In response to the problems arising from the usage of surfactants, an alternative pathway to synthesise ASE without surfactant was developed. In 2011, Suau patented such a synthesis route that uses a co-monomer 2-acrylamido-2-methylpropane sulfonic acid, or AMPS for short, on top of MAA and EA.[38] According to his protocol, the reactor has a volume of 1 liter and is equipped with a mechanical stirrer and oil bath. Firstly, 605 g of bipermutated water was mixed with 7.5 g of AMPS solution (1.2 wt%) and heated to 82 °C. Then, a catalyst that is made up of 1 g of APS and 0.1 g of sodium metabisulfite dissolved in 20 g of bipermutated water was added through a funnel. Afterwards, the monomers (36.5 wt% MAA and 62.3 wt% EA) were added continuously and gradually. The temperature of the medium was maintained at 85 °C throughout the reaction. After the monomers were added, the pump was rinsed with 15 g of bipermutated water and allowed to further react for half an hour, keeping the temperature constant. The final step would be the addition of 0.15 g of APS solution diluted in 20 g of water for another half an

Table 3.2 Ingredients for P(MAA-EA) synthesis.

Ingredients	Function	Amount
Ethyl acetate (EA)	Monomer	16.3 g
Methacrylic acid (MAA)	Monomer	0.04 g
Diallyl phtalate	Cross-linker	0.04 mL
Sodium dodecylsulfate (SDS)	Surfactant	0.22 g
Ammonium persulfate (APS)	Free radical initiator	0.1 g
Water	Solvent	51 g

hour at the same temperature. The temperature was raised by 2 °C and the reaction was allowed to continue for an hour.

The final reaction product was a perfectly homogeneous emulsion. The solids content of the emulsion was found to be high (34 wt%) and the particle size, as characterised by dynamic light scattering, was 170 nm. The emulsion was then dispersed in water so that the active ingredient content of 5% was achieved. The medium was then neutralised with NaOH to a pH of 6.5. The viscosity of the medium as measured by a Brookfield viscometer at 25 °C and 100 rev min^{-1} was 3420 mPa s.

On another test, a similar protocol was replicated, but there was no addition of AMPS. The result yielded a non-homogeneous cloudy precipitate in the solution having a particle size of 400 nm, as characterised by dynamic light scattering. One problem foreseen with such a large particle size is the sedimentation of the product upon storage. The amount of AMPS added should also not be too high. It was found that adding 25 wt% of AMPS also formed an insoluble compound that readily precipitated into the medium. Similarly, when AMPS was not added or was replaced with another co-monomer like styrene, lauryl methacrylate, sodium styrene sulfonate, *etc.*, the precipitates were formed.[38]

3.5 Hydrophobically Modified Alkali Swellable Emulsion (HASE)

Hydrophobically modified alkali swellable emulsion (HASE) is considered to be a hybrid thickener which increases the viscosity of cosmetic products by both hydrodynamic thickening and hydrophobic interactions. HASE thickener is synthesised by adding ethoxylated long chain hydrophobic moieties to the ASE thickener polymer backbone, forming a terpolymer. The hydrophobic moieties are typically (meth)acrylic ester of an ethoxylated long chain alcohol. One example of such hydrophobic moieties is VISIOMER® C18 PEG 1105 MA W, a readily available product supplied by EVONIK. The structure of such a compound is shown in Figure 3.8. Figure 3.9 shows the possible chemical structure of HASE.[34]

Similar to ASE, HASE is supplied at low pH. Upon neutralisation with base, the polymers are negatively charged. The charge induced chain extension and the association of the hydrophobic groups causes the viscosity of the solution to increase.[1] This hydrophobic group association can be imagined

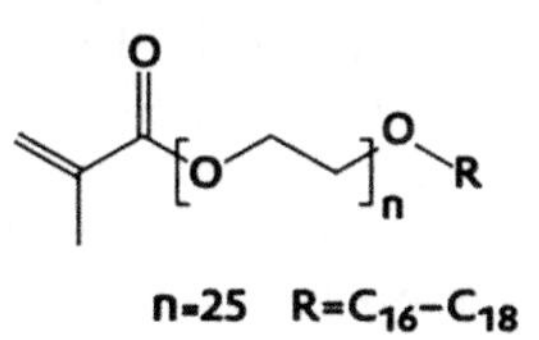

Figure 3.8 Example of a hydrophobic moiety VISIOMER®C18 PEG 1105 MA W.

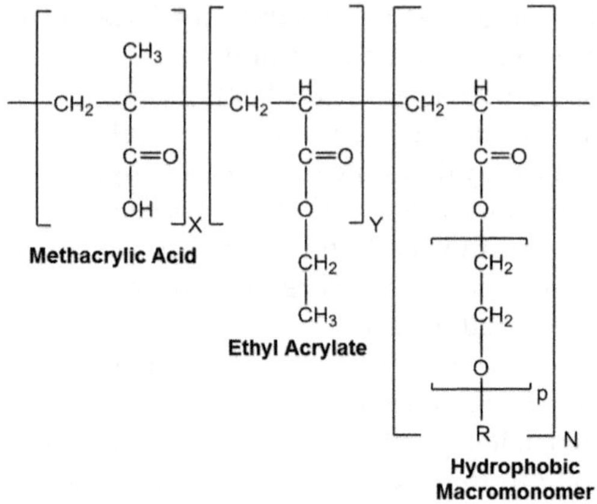

Figure 3.9 Structure of common HASE with the hydrophobic moiety incorporated.

in a similar way to how the micelles are formed from surfactants, where the hydrophobic groups aggregate together in the water phase.[13]

To synthesise HASE, Abdel-Wahhab *et al.* reported a method using a 2 liter reactor equipped with a magnetic stirrer, heater and nitrogen gas inlet. Afterwards, 1.66 mol of distilled water were mixed with the surfactant sodium lauryl ether sulfate (SLES, 68% concentration) and heated up to a temperature of 85 °C. A pre-emulsion could be prepared by mixing 30 g of monomer mixture (of EA, MAA and C18PM), 0.9 g of SLES and 1.61 mol of distilled water in the flask. The pre-emulsion was stirred at 500 rpm for 1 hour. The reaction could be initiated by dissolving 0.0008 mol of sodium persulfate in 0.54 mol distilled water. The temperature remained the same at 300 °C but the stirring rate was reduced to 200 rpm. The pre-emulsion and initiator were added incessantly and gradually for 4 hours, keeping the temperature as close to 85 °C as possible. After the addition was complete, the flask was allowed to stir for another 2 hours to ensure a high monomer conversion.[26]

To study the effect of concentration of hydrophobic moiety (C18PM) on the viscosity of HASE, different polymers were synthesised using the above method, varying the composition of EA and C18PM, while keeping the MAA concentration constant at 40%. The products were cooled, filtered and diluted in distilled water into polymer solution with 1% solids content. The viscosity was measured by LVD V-II Brookfield viscometer at 25 °C after adjusting the pH to 7.5 with aqueous NH_4OH solution. It was shown that the viscosity increases as the concentration of the hydrophobic moiety increases from 0 to 8%. Beyond 8%, the viscosity started to decrease and beyond 10%, precipitates began to form. In the absence of the hydrophobic group, the polymer practically becomes ASE and the viscosity is only 100 mPa s. The thickening effect is caused only by the electrostatic repulsion between $-COO^-$ in methacrylic acid that forms hydrogen bonding and water. In the presence of C18PM, the thickening effect is amplified by the hydrophobic association.[26] Refer to Figure 3.10 for a graphical representation of this thickening effect due to the hydrophobic groups.

The investigation was further explored to study the effect of the EA to MMA ratio on the thickening of the product. From the previous results, the concentration of the hydrophobic moiety C18PM was kept constant at the optimum level of 8%. The EA:MMA:C18PM ratio that gave maximum thickening effect was 62:30:8, as highlighted in Figures 3.11 and 3.12. When the concentration of MMA increased beyond 40%, the precipitates began to form largely.[26]

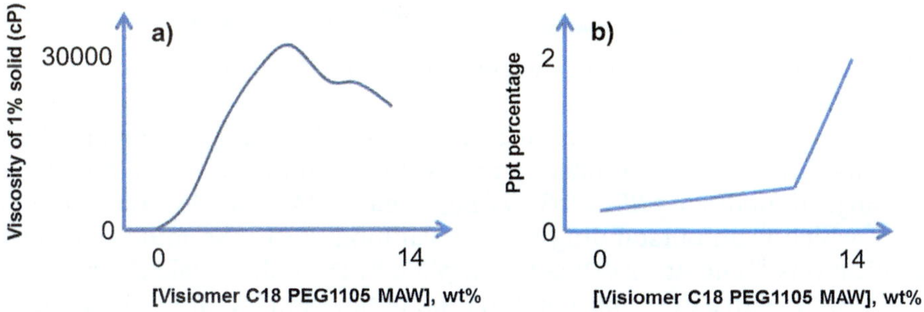

Figure 3.10 (a) Effect of concentration of hydrophobic groups on the viscosity of HASE.[26] (b) Effect of concentration of hydrophobic groups on the precipitates formed.[26]

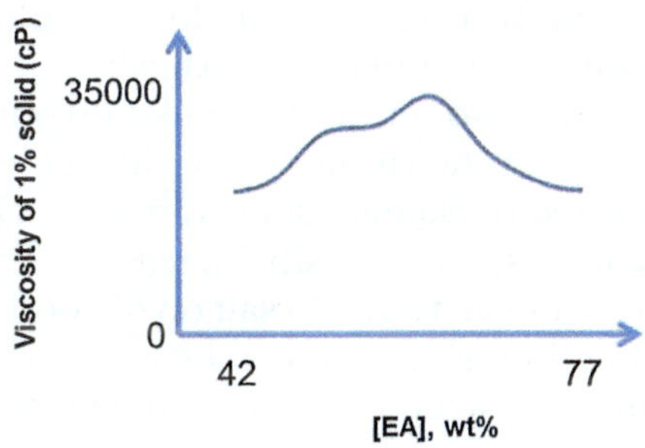

Figure 3.11 Effect of EA:MMA ratio on the viscosity of HASE.[26]

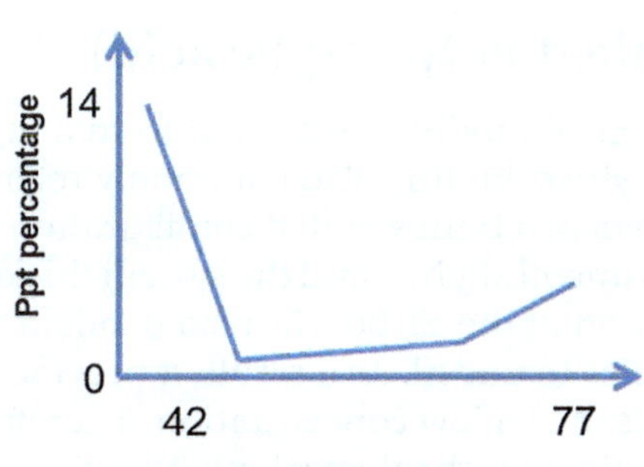

Figure 3.12 Effect of EA:MMA ratio on the precipitates formed.[26]

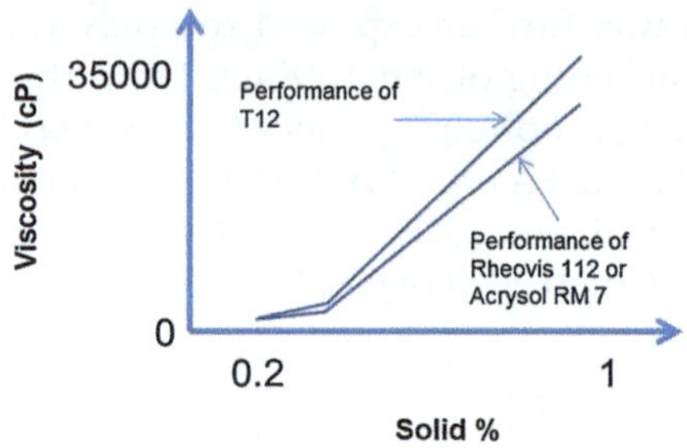

Figure 3.13 Effect of solids content on viscosity of different HASEs.[26]

The viscosity of this optimum composition of HASE (labelled as T12 by the experimenter) was compared against two commercially available HASE rheological modifiers: *Rheovis® 112* supplied by BASF at 40% solids and pH 3.2,[39] which is an outstanding rheological modifier in the mid-shear range applications,[40] and *Acrysol™ RM-7* supplied by Dow Chemical, which is a versatile HASE with a desirable combination of flow and levelling when used as paint.[41] T12 gives a better thickening effect in water than Rheovis® 112 and Acrysol™ RM-7 at all concentrations, as seen in Figure 3.13. It can therefore be used to thicken an aqueous base medium at low concentration. T12 also had a higher electrolyte resistance in NaCl than the other two HASEs. Consequently, it can be suitably used in cosmetics formulations which contain electrolytes, such as shampoos and liquid soaps.[42]

The experiments above also tell us the effect on solids content and the presence of salt solution on the viscosity of a rheological modifier formulation. In general, the three thickeners show the same characteristics. When the solids content increases, the viscosity increases. The increase in concentration of ionic electrolytes such as NaCl salt on the other hand decreases the viscosity of the formulation. NaCl salt can be classified as "water-structure makers" which enhance the structure of the water when added to a solution. The hydrophobic effect would as a result be enhanced, reducing the viscosity of the formulation.

3.6 Cross-Linked Poly(acrylic acid)

This group of rheological modifier is derived from a high molecular weight homopolymer of acrylic acid. It is also commonly referred to as carbomer.[4] When it is dry, it exists in a tightly coiled configuration. As it is dispersed in water, the molecules uncoil slightly and the system thickens to a small extent. However, neutralisation of the carboxylic acid pendant groups increases the ability of the molecules to uncoil. As a result, it provides a dramatic thickening even when it is used at a low concentration. Therefore, carbomer can be considered as a very efficient rheological modifier.[15]

One method to prepare poly(acrylic acid) is by reversible addition–fragmentation transfer (RAFT) polymerisation. Firstly, 0.199 mol of acrylic acid and 0.00186 mol of trithiocarbonic acid dibenzyl ester as the chain transfer agent in 0.876 mol of ethanol are mixed in a 100 mL flask equipped with a cold water condenser. Then, the mixture is degassed with nitrogen. The temperature of the reaction is raised to 80 °C. The polymerisation is initiated by adding 1.32×10^{-4} mol 4,4′-azobis(4-cyanovaleric acid) inhibitor. An aqueous solution of 10 M NaOH is added in excess to neutralise PAA and unreacted monomer and to cleave the basic (pH > 9) dithio-terminated chains. Finally, this mixture is further stirred for 2 hours and then dried *in vacuo*.[43] Afterwards, carbomer can be synthesised by cross-linking poly(acrylic acid) by allyl ether of pentaerythritol, allyl ether of sucrose or allyl ether of propylene.[15]

For many applications, poly(acrylic acid) is also used in the form of its salt, sodium polyacrylate. The thickening mechanism of sodium polyacrylate is similar to the thickening mechanism of poly(acrylic acid), in which water molecules enter the polymer hydrogel matrix with subsequent swelling. The swelling of the dispersed polymers provides an elegant feel on skin, thus it finds many applications in cosmetic products.[44]

The effect of pH, temperature and polymer concentration on the viscosity of sodium polyacrylate thickener was studied by Medina-Torres *et al.*[44] These three factors were thought to be the most important factors affecting the viscosity of sodium polyacrylate in cosmetic application. The experiments were carried out by varying one factor while keeping the other factors constant. The measurements of the viscosity were performed on an AG2 (model TA) controlled stress rheometer. The concentric cylinders used were 25 mm in diameter and the diameter of the parallel plates was 25 mm.

Firstly, to evaluate the effect of concentration, the viscosities of solutions of different concentrations of sodium polyacrylate in emulsion (1.0%, 1.5% and 2.0%) were measured at $T = 25$ °C and pH = 6.5. Secondly, to comprehend the effect of temperature, the viscosities of solutions having a 2% concentration of sodium polyacrylate in emulsion were measured at a constant pH of 6.5 but different temperatures (10 °C, 25 °C and 45 °C). Lastly, the effect of pH was investigated by varying the pH to 4.0, 6.5 and 9.0 at 25 °C and 2% sodium polyacrylate concentration.[44]

One general conclusion is that the sodium polyacrylate displays shear thinning (pseudoplastic) behaviour, as the viscosity decreases with increasing shear rate. When the shear rate becomes higher, the fluid structure is disrupted and the polymer molecules align themselves towards the direction of the flow, reducing its viscosity. Stronger interaction between a higher interphase area and more dispersed particles is produced at higher shear rates.[44]

All samples possess similar shear-thinning properties. The viscosity increases with sodium polyacrylate content in the emulsion. The increase in viscosity is more pronounced at lower concentration (from 1.0% to 1.5%). Above 1.5% concentration, the viscosity becomes stabilised. In the 1.5–2.0% concentration range, the swelling condition for sodium polyacrylate is optimised, explaining the small increment in viscosity when the concentration was increased from 1.5% to 2.0%.[45,46]

The increase in temperature indeed increases the viscosity of the solution. However, the increase in viscosity is rather negligible. Together with the fact that the slopes of the three curves are fairly similar, it can be concluded that the flow consistency index at different temperatures is constant and thus sodium polyacrylate is stable in the temperature range of 10–45 °C (skin surface temperature from summer to winter conditions). Similar findings were obtained at different concentrations of sodium polyacrylate.[44] The change in viscosity does not contribute to the macrostructural changes in the flow behaviour at elevated temperatures.[47]

Last but not least, the low pH of the formulation reduces the viscosity. The slope of the curve at pH = 4.0 is also less steep compared to the other two at higher pH. This indicates that the polymer has almost Newtonian behaviour at low pH. As the pH decreases, the cation–anion interaction is reduced, causing a loss of structure in the emulsion. Furthermore, the particle–particle interactions are reduced and the polymer chain now has a collapsed structure. At a pH above 6, a stable network is established and particle–particle interactions are promoted.[45] This thickening effect is stable up to pH 9, indicating that sodium polyacrylate is suitable for a wide range of personal care applications.[44]

3.7 Safety Issues with Polyacrylates

Long term toxicity of polyacrylate used in consumer care products has not been strongly established, even though in 2009 there was the first case reported by Song *et al.* related to clinical toxicity of polyacrylate nanoparticles in humans.[48] They argued that serious damage to human lungs can surface due to prolonged exposure to polyacrylate nanoparticles found in the workplace as well as in emulsion formulations in adhesives and paint. The conclusions were drawn from the pathological examination of the lung tissues of the patients and characterisation using transmission electron microscopy (TEM) which observed round particles, ~30 nm in diameter, in chest fluid as well as in the cytoplasm and karyoplasms of epithelial and mesothelial cells in the pulmonary system. However, Ren and Huang a year later pointed out the flaw of the argument in their paper. In fact, this false reporting attracted the attention of scientists worldwide to gather at the China International Conference on Nanoscience and Technology to evaluate the credibility of the claim.[49]

According to Ren and Huang, the data presented by Song *et al.* are not sufficient to substantiate that the clinical symptoms in the patients were solely attributed to the toxicity of the polyacrylate nanoparticles. Adequate experimental evidence based on *in vivo* and *in vitro* assessments should be provided to substantiate their argument. Moreover, some other nanoparticles like thin film zinc oxide, titanium dioxide and silicon nanoparticles are commonly added into the polyacrylate emulsion to increase the strength and abrasion resistance of the material.[48,50–53] It is therefore suspected that these particles may be the real cause of the lung damage, as the report by Song *et al.* was not substantiated by the compositional analysis of the nanoparticles mentioned. However, it should not be implied that polyacrylates are not toxic. Further studies including animal testing, *in vivo* and *in vitro* experiments should be carried out to confirm this hypothesis.

Despite being regarded as having low toxicity, acrylate copolymers (*e.g.* ASE/HASE) containing residual monomers of 2-ethylhexyl acrylate are known to produce irritation. Nonetheless, there is no evidence of sensitisation related to them. 2-Ethylhexyl acrylate is not genotoxic, but it is carcinogenic at a concentration higher than 21% tested on the skin of mice. Furthermore, dermal and nasal irritations also occured in this study. However, the concentration of monomers used in cosmetic formulations is usually lower than the irritation threshold for humans, thus it is not considered to exhibit a safety risk. As an exception, ethyl methacrylate, a common formulation in nail polishes, was found to cause sensitising when it is applied to skin. However, this problem can be minimised by presenting a warning in the packaging of the product.[6] Additionally, dilute sodium polyacrylate solutions cause eye irritations in rabbits and general respiratory collapse and widespread tissue damage upon injection, but they do not irritate the skin of rabbits and have low acute oral toxicity.[54] These results need an extrapolation to the human body counterpart in order for us to make a thorough conclusion of the overall safety issues with polyacrylates.

A few polyacrylates were tested for impurities with a gas chromatography method. The result shows that the residual monomers present in the sample are within the safety level or even below the detection limit. For example, the residual amount of acrylic acid is <0.2 to 0.8 ppm, 0.08 to 2.6 ppm of methyl methacrylate and 1.3 to 3.9 ppm ethylene glycol dimethacrylate.[55] In poly(acrylic acid), the residual acrylic acid detected by UV spectroscopy at 195 nm was 300 ppm.[6] This further substantiates the low toxicity level of polyacrylates.

3.8 Comparing Different Polyacrylates

It is clear that adding the hydrophobic moiety to ASE causes the viscosity of the rheological modifier to increase multiple times, as there are more sites of interaction that can form hydrophobic associations in aqueous medium.[26] Thus, HASE has generally higher viscosity than ASE and is used in cosmetics formulations that require higher viscosity. Practically, HASE becomes more largely accepted as a rheological modifier than ASE, mainly because of its technical performance such as good levelling and flowing that can be adjusted by choosing the most suitable hydrophobic moiety. Moreover, HASE is also price competitive. The rheological characteristics of HASE are more Newtonian and less shear thinning than ASE.[13]

One major advantage of using sodium polyacrylate as a rheological modifier is that there is no neutralisation needed before application. On the contrary, ASE and HASE thickeners must be neutralised by pH adjustment to 8.0–9.5 before they are used, so that the thickening efficiency is maximised and the viscosity is stably maintained. The use of sodium polyacrylate simplifies the preparation of cosmetic products and increases the potential applications of sodium polyacrylates.[44]

3.9 Comparing Polyacrylates with Other Rheological Modifiers

The tabular evaluation of using polyacrylates in comparison to other types of rheological modifiers, namely cellulosic, associative and specialty clays, is summarised in Table 3.3. The comparison of polyacrylates with associative thickeners like polyurethane and polyethers as a rheological modifier is presented in Table 3.4. Basically, polyurethane and polyester thickeners are pH independent and have little effect on water sensitivity. They have a lower molecular weight to reduce spattering and higher number of hydrophobic groups per molecular volume than polyacrylates.[34]

3.10 Outlook, Perspectives and Recommendations

Polyacrylate polymers generally show good performance as rheological modifiers in cosmetics formulations, as they have a strong shear thinning ability. The viscosity of the formulations can also be fine-tuned by a number of factors, for example, an addition of hydrophobic moieties and an increase in concentration will increase the viscosity.

Table 3.3 Evaluation of using different types of rheological modifiers.

Product	Advantages	Limitations
Cellulosic	Shear thinning for easy application	Flow and levelling
	Colorant compatibility	Roller spattering
	Open time control by water retention	Effect on water and scrub resistance
	Sag control	Negative influence on gloss
		Easy biodegradation
Acrylics (ASE, HASE)	Strong shear thinning	pH sensitivity
	Anti-setting and anti-sag	
	Cost advantage	Effect on water and scrub resistance
	Good spray properties	
Associative (HEUR, HMPE)	Excellent flow and levelling	Viscosity-loss-on-tinting
	Low shear thinning	Sag control
	Minimized roller spattering	
	Gloss	
	High film build	
Speciality clays	Sag resistance	Incorporation
	Heat resistance	Flow and levelling
		Open time control

Table 3.4 Comparison of different rheological modifiers for a water-based system.

Property/influence on	Polyurethanes, polyethers	Acrylics
Low shear viscosity	Medium contribution	High contribution
High shear viscosity	High contribution (high film thickness by roller/brush application possible)	Low contribution, less impact on film thickness
Spattering	Improves spattering	Low effect
Gloss	Low to no effect	Matting effect possible
Influence of surfactants	Low to high, must be tested in each case	Hardly to moderate
Syneresis	Tendency to syneresis	Low to no tendency to syneresis depending on type
Scrub and wash resistance	Very good effect	Good to moderate effect depending on type

The rheological properties must be assessed in the conditions that mimic the site where the personal care products are going to be used. Let us take an example of acrylates copolymer rheological modifiers used in a skin care formulation. Normally, they would be supplied as a white mobile liquid having pH 3.0 at room temperature.[6] When they are applied to the skin, the local temperature would increase to about 37 °C and therefore it is important to assess whether or not the rheological properties of the polymer would change when subjected to a higher temperature. If the viscosity decreases significantly, a higher concentration of the polymer can be added to the formulation to compensate for the decrease in viscosity.[26]

In response to the good manufacturing practice (GMP) principle, a manufacturer should not produce products that harm the end user.[56] Even though polyacrylates have a generally low toxicity level, it should still be noted that some components of rheological modifiers (*e.g.* the residual monomers) and other formulations of cosmetic products may trigger cytotoxicity, inflammation, sensitisation or even gene mutation in the human body.[6] Currently, research on cosmetic product safety is still limited to *in vitro* studies and *in vivo* in animals.[49] Tissue-engineered skin models are widely developed to mimic the *in vivo* conditions and test whether irritation, sensitisation or cytotoxicity would happen.[57–59] It is advisable that all personal care product manufacturers take an extra step to expand their clinical testing to *in vivo* tests in humans, *ex vivo* tests and test runs on human volunteers. As such, they can validate the claim of the effectiveness of their cosmetic products and thoroughly monitor the side effects of their products.[56]

Furthermore, in some more serious cases, although the materials that make up the formulations are clinically proven to be safe, the synthesis method (*i.e.* sterilisation with ionising radiation) could trigger the formation of free radicals.[60] These residues are harmful and might cause the materials to fail *in vivo* as they can bind actively to components of the human body.

Therefore, it is advisable to choose a suitable sterilisation method which does not trigger the formation of free radicals during the manufacturing process. Otherwise, free radical scavengers[61] or antioxidants[62] can be introduced to the formulation in the last step of the manufacturing process in order to eliminate the free radicals generated by synthesis process.

The use of different formulations in personal care products with respect to ecotoxicity (environmental toxicity) has become a global concern. For example, since June 2007 the European REACH regulation programme has aimed to improve human health and environmental protection by offering better information about chemical properties (ecotoxicity) and improving the spread of information to ensure there are no safety issues with these chemicals.[63] In response to this global concern, much research has been done to develop products that have low ecotoxicity levels. It has been found that grafting silica nanoparticles of size 22 nm to HASE polymers results in a smaller degree of ecotoxicity than the non-grafted polymers.[64] Moreover, grafting nanoparticles to a HASE system also gives additional benefits like homogeneous dispersion of nanoparticles and lower biological toxicity.[65] Hence, nanoparticle-stabilised HASE definitely has a huge potential to be utilised in cosmetic formulations with regard to the global movement to use greener products.

Polyacrylate nanoparticles also exhibit potential further advantages in nanomedicine. They can act as carriers for drug delivery, pass across physiological barriers and target specific organs and cells.[66–69] With advancements in the technology, it is highly possible to incorporate drugs into polyacrylate used in cosmetic formulations to treat not only skin-related diseases but also various other diseases. It might seem a little far-fetched now, but given enough time, I believe polyacrylates have never-ending potential in personal care products and the nanomedicine field.

References

1. A. Patil and M. S. Ferritto, *ACS Symp. Ser.*, 2013, 1148, 3–11.
2. M. D. L. d. Castro and A. Townshend, in *Analysis of Cosmetic Products*, ed. A. S. C. S. C. T. S. Carreño, Elsevier, Amsterdam, 2007, pp. xi–xii, DOI: 10.1016/B978-044452260-3/50021-8.
3. J. V. Gruber, *Principles of Polymer Science and Technology in Cosmetics and Personal Care*, Marcel Goddard, Inc., New York, 1999.
4. R. Y. Lochhead, *ACS Symp. Ser.*, 2007, 961, 3–56.
5. P. A. Williams, *Handbook of Industrial Water Soluble Polymers*, Blackwell Publishing Ltd, 2007, pp. 1–9, DOI: 10.1002/9780470988701.ch1.
6. M. Zondlo Fiume, *Int. J. Toxicol.*, 2002, **3**(**21 Suppl**), 1–50.
7. J. A. Wenninger, G. N. McEwen, R. C. Canterbery, T. Cosmetic and F. Association, *International Cosmetic Ingredient Dictionary and Handbook*, Cosmetic, Toiletry, and Fragrance Association, 1999.
8. M. Hawe, *Handbook of Industrial Water Soluble Polymers*, Blackwell Publishing Ltd, 2007, pp. 32–72, DOI: 10.1002/9780470988701.ch3.

9. R. Y. Lochhead, *ACS Symp. Ser.*, 2010, 1053, 3–22.
10. T. C. Inc., *Estimate of global production capacity for polyacrylamides/ polyacrylates in 2004*, 2004.
11. A. Kamal, M. S. Kumar, C. G. Kumar and T. Shaik, *J. Microbiol. Biotechnol.*, 2011, **21**, 37–42.
12. L. Gilbert, G. Savary, M. Grisel and C. Picard, *Chemom. Intell. Lab. Syst.*, 2013, **124**, 21–31.
13. BASF, *Little Helpers Love Great Achievement: Practical Guide to Rheology Modifiers*, BASF, Ludwigshafen, 2013.
14. A. O. Barel and H. I. Maibach, *Handbook of Cosmetic Science and Technology*, CRC Press, 2001.
15. D. D. Braun and M. R. Rosen, *Rheology Modifiers Handbook: Practical Use and Applilcation*, Elsevier Science, 2008.
16. P. Sherman, *Fette, Seifen, Anstrichm.*, 1964, **66**, 370.
17. F. W. Billmeyer, *Textbook of Polymer Science*, Wiley, 1984.
18. J. R. Fried, *Polymer Science and Technology*, Prentice Hall PTR, 1995.
19. R. Brummer, *Rheology Essentials of Cosmetic and Food Emulsions*, Springer Berlin Heidelberg, 2006.
20. A. A. Collyer and D. W. Clegg, *Rheological measurement*, Elsevier Applied Science, 1988.
21. J. V. Lawler, *Polym. Adv. Technol.*, 1994, **5**, 620.
22. B. J. Berne and R. Pecora, *Dynamic Light Scattering: With Applications to Chemistry, Biology, and Physics*, Dover Publications, 2013.
23. C. M. Miller, E. D. Sudol, C. A. Silebi and M. S. El-Aasser, *J. Colloid Interface Sci.*, 1995, **172**, 249–256.
24. S. T. Balke, A. E. Hamielec, B. P. LeClair and S. L. Pearce, *Prod. R&D*, 1969, **8**, 54–57.
25. G. Gellerstedt, in *Methods in Lignin Chemistry*, ed. S. Lin and C. Dence, Springer Berlin Heidelberg, 1992, ch. 34, pp. 487–497, DOI: 10.1007/978-3-642-74065-7_34.
26. H. Abdel-Wahhab, A. Naser, F. Abdel-Hai, S. I. Mohamed and A. Abdel-Hakim, *J. Appl. Sci. Res.*, 2011, **7**, 1386–1394.
27. W. Bauer, *Kirk-Othmer Encyclopedia of Chemical Technology*, John Wiley & Sons, Inc., 2000, DOI: 10.1002/0471238961.0103182502012105.a01.pub2.
28. C. E. Habermann, *Kirk-Othmer Encyclopedia of Chemical Technology*, John Wiley & Sons, Inc., 2000, DOI: 10.1002/0471238961.0103182508010205.a01.pub2.
29. H. Erpenbach, K. Gehrmann, H. Handte, H. Reis and K. Sennewald, *US Pat.*, US3553261, 1971.
30. J. J. Kurland and D. R. Bryant, *Plant/Oper. Prog.*, 1987, **6**, 203–207.
31. H. Rauch-Puntigam and T. Völker, *Acryl- und Methacrylverbindungen*, Springer, 1967.
32. M. Salkind, E. H. Riddle and R. W. Keefer, *Ind. Eng. Chem.*, 1959, **51**, 1328–1334.
33. W. Bauer, *Ullmann's Encyclopedia of Industrial Chemistry*, Wiley-VCH Verlag GmbH & Co. KGaA, 2000, DOI: 10.1002/14356007.a16_441.

34. *Synthetic Thickeners 101 – Alkali Swellable Thickeners*, http://www.snpinc.com/resources/alkali-swellable-thickeners/.
35. J. K. Oh, L. Deleebeeck, M. A. Winnik, J. Rademacher and R. Farwaha, *J. Polym. Sci., Part A: Polym. Chem.*, 2005, **43**, 5632–5642.
36. A. M. van Herk, S. A. F. Bon and K. Landfester, *Hybrid Latex Particles: Preparation with (Mini)emulsion Polymerization, Springer*, 2010.
37. G. Odian, *Principles of Polymerization*, John Wiley & Sons, Inc., 2004, pp. 350–371, DOI: 10.1002/047147875X.ch4.
38. J. M. Suau, Google Patents, 2011.
39. BASF, *Little Helpers Love Great Achievement: Formulation Additives by BASF*, Ludwigshafen, 2013.
40. BASF, *Technical Information: Rheovis® HS 1212 (old : Rheovis® 112)*, BASF, 2013.
41. T. D. C. Company, *ACRYSOL™ RM-7: Rheology Modifier For Waterborne Architectural and Industrial Coatings*, The Dow Chemical Company, 2014.
42. R. Y. Lochhead, *Polymers as Rheology Modifiers*, American Chemical Society, 1991, ch. 6, vol. 462, pp. 101–120.
43. J. Loiseau, N. Doërr, J. M. Suau, J. B. Egraz, M. F. Llauro, C. Ladavière and J. Claverie, *Macromolecules*, 2003, **36**, 3066–3077.
44. L. Medina-Torres, F. Calderas, G. Sanchez-Olivares and D.-M. Nuñez-Ramirez, *Ind. Eng. Chem. Res.*, 2014, **53**, 18346–18351.
45. O. Okay and S. B. Sariisik, *Eur. Polym. J.*, 2000, **36**, 393–399.
46. T. Tanaka, *Polymer*, 1979, **20**, 1404–1412.
47. M. Siewert, J. Dressman, C. K. Brown and V. P. Shah, *AAPS PharmSciTech*, 2003, **4**, E7.
48. Y. Song, X. Li and X. Du, *Eur. Respir. J.*, 2009, **34**, 559–567.
49. H. Ren and X. Huang, *Eur. Respir. J.*, 2010, **36**, 218–221.
50. T. Mizutani, K. Arai, M. Miyamoto and Y. Kimura, *Prog. Org. Coat.*, 2006, **55**, 276–283.
51. H. A. Ali and A. A. Iliadis, *Thin Solid Films*, 2005, **471**, 154–158.
52. Y. Chen, A. Lin and F. Gan, *Appl. Surf. Sci.*, 2006, **252**, 8635–8640.
53. A. Mitra and A. Bhaumik, *Mater. Lett.*, 2007, **61**, 659–662.
54. Bibra, *BIBRA*, 1987.
55. C. P. L. Inc., *Residual monomers test for Daitosol 5000AD (Acrylates Copolymer)*, 1996.
56. K. Lintner, *Global Regulatory Issues for the Cosmetics Industry*, Elsevier Science, 2009.
57. T. Welss, D. A. Basketter and K. R. Schröder, *Toxicol. In Vitro*, 2004, **18**, 231–243.
58. A. Roy, R. K. Sahu, M. Matlam, V. K. Deshmukh, J. Dwivedi and A. K. Jha, *Pharmacogn. Rev.*, 2013, **7**, 97–106.
59. K. Schlotmann, M. Kaeten, A. F. Black, O. Damour, M. Waldmann-Laue and T. Forster, *Int. J. Cosmet. Sci.*, 2001, **23**, 309–318.
60. K. A. Athanasiou, G. G. Niederauer and C. M. Agrawal, *Biomaterials*, 1996, **17**, 93–102.

61. T. R. Green, S. R. Bennett and V. M. Nelson, *Toxicol. Appl. Pharmacol.*, 1994, **129**, 163–169.
62. J.-M. Lü, P. H. Lin, Q. Yao and C. Chen, *J. Cell. Mol. Med.*, 2010, **14**, 840–860.
63. E. P. a. C. R. (EC), *Off. J. Eur. Union*, 2007, 361.
64. L. Clément, A. Zenerino, C. Hurel, S. Amigoni, E. Taffin de Givenchy, F. Guittard and N. Marmier, *Sci. Total Environ.*, 2013, **450–451**, 120–128.
65. L. Lacroix-Orio, *SATT SUD-EST*, 2014.
66. J. C. Sung, B. L. Pulliam and D. A. Edwards, *Trends Biotechnol.*, 2007, **25**, 563–570.
67. H. Otsuka, Y. Nagasaki and K. Kataoka, *Adv. Drug Delivery Rev.*, 2003, **55**, 403–419.
68. E. Turos, J.-Y. Shim, Y. Wang, K. Greenhalgh, G. S. K. Reddy, S. Dickey and D. V. Lim, *Bioorg. Med. Chem. Lett.*, 2007, **17**, 53–56.
69. E. Turos, G. S. Reddy, K. Greenhalgh, P. Ramaraju, S. C. Abeylath, S. Jang, S. Dickey and D. V. Lim, *Bioorg. Med. Chem. Lett.*, 2007, **17**, 3468–3472.
70. FDA, *Frequency of use of cosmetic ingredients*, Washington, DC, 1998b.

CHAPTER 4

Natural Rheological Modifiers for Personal Care

YUJIE JASON ZHENG[b] AND XIAN JUN LOH*[a,b]

[a]Institute of Materials Research and Engineering, A*STAR (Agency for Science, Technology and Research), 2 Fusionopolis Way, Innovis, #08-03, Singapore 138634, Singapore; [b]Department of Materials Science and Engineering, National University of Singapore, Singapore 117574, Singapore
*E-mail: lohxj@imre.a-star.edu.sg

4.1 Introduction

Personal care products can be found in almost any household nowadays as they are ubiquitous consumer goods that have been regarded as a staple since ancient times. Personal care products, otherwise known as cosmetics, are not a modern invention and can be traced back to as early as the Egyptian, Roman and Greek eras.[1] Historically, Egyptian women used to apply kohl to darken their eyelids while Greek women used makeup containing poisonous lead carbonate to achieve a whiter complexion, resulting in unfortunate loss of lives.[2] It is claimed that the Egyptian queen, Cleopatra bathed in donkey's milk so as to keep her skin smooth and supple.[2,3] In Roman times, bathing was also a prominent activity, evident from dated records which indicate that a soap factory was found in the ruins of Pompeii, a city destroyed by the Mount Vesuvius eruption in 79 A.D.[4] Hence, the social values and ideologies of upholding one's personal hygiene and grooming, as well as beautifying oneself, were already established then.

RSC Polymer Chemistry Series No. 20
Polymers for Personal Care Products and Cosmetics
Edited by Xian Jun Loh

Published by the Royal Society of Chemistry, www.rsc.org

According to the US Food and Drug Administration (FDA), it is generally understood under the Federal Food, Drug, and Cosmetic Act (FD&C Act) that cosmetics are "articles intended to be rubbed, poured, sprinkled, or sprayed on, introduced into, or otherwise applied to the human body for cleansing, beautifying, promoting attractiveness, or altering the appearance".[5] Personal care products can be treated as a broadly-used term to refer to any form of substance that is used for the purpose of personal grooming and hygiene, beautifying our appearance or to alter our odours. The diverse range of personal care includes products such as skincare lotions, makeup, soaps, shower gels, shampoos, conditioners, perfumes, deodorants, as well as toothpastes, mouth washes and many more. The importance of these commodity products should not be underestimated as they uphold a pivotal role entrenched in our daily lives by increasing our personal health and well-being as well as enhancing our self-confidence and feelings of attractiveness.

A case in point is hand washing with soap in particular, which has been identified as the most cost-effective measure for disease control among various health promotion campaigns.[6–8] Studies have shown that hand washing could save more than a million lives annually from diarrhoeal diseases and respiratory infections, which are two of the top causes of child mortality in developing countries.[6,9–11] Even in developed countries, hand washing could prevent the spread of infectious viruses, such as norovirus, rotavirus and influenza and hospital-acquired infections, including methicillin-resistant *Staphylococcus aureus* and *Clostridium difficile*.[6,12–16] A recent UK study even revealed that a quarter of sampled commuters had faecal bacteria on their hands.[6,17]

Another worthy case to consider is that even today, society still upholds a strong emphasis on personal image and appearance due to the continuous pursuit of beauty. This could be attributed partly to the elusive nature of beauty, which is a fluctuating ideal that varies across cultures and over time.[18,19] However, definition dictates that an 'ideal' can only be achieved by a minority of those who strive for it. One of the reasons why beauty ideals change constantly is because the value of beauty standards depends distinctly on its extraordinary nature that can only be attained by a small proportion of the population. The growing pursuit of beauty has normalised the significance of attractiveness in our culture, such that serious ramifications could result for social transgressions. Attractiveness forms a major part of one's first impression as well as in subsequent interpersonal relationships. In a classical study, "What Is Beautiful Is Good," psychologists Kenneth Dion, Ellen Berscheid and Elaine Hatfield asked college students to rate photographs of strangers on a variety of personal characteristics. It was found that those that were judged to be attractive were also more likely to be rated positively as intelligent, kind, happy, flexible, interesting, confident, sexy, assertive, strong, outgoing, friendly, poised, modest, candid and successful than those judged unattractive.[20] Furthermore, unattractiveness could even lead to real consequences, such as the limitation of career progression and opportunities[21,22] or finding accommodation.[23]

4.1.1 Personal Care Market and Trends

The market for personal care products encompasses 6 main segments: (1) skincare, (2) haircare, (3) colour (makeup), (4) fragrances, (5) toiletries (soap)/deodorants and (6) oral care. Skincare consists of an entire range of washing products, including soaps, perfumed soaps, shower gels and lotions. Haircare includes a whole range of treatments, including shampoos and conditioners, and then branching out to including a multitude of dye products from full colour treatments to highlights. Oral care has expanded into the realm of cosmetic surgery; besides conventional toothpastes there are also mouth washes, tooth polishes and whitening pastes.[4]

From 2011–2013 statistics, skincare products made up the largest proportion of the global market at 31%.[24] The industry remains a viable growing sector according to market researcher Lucintel; the global personal care products industry is predicted to reach $630 billion by 2017, with a CAGR of 3.4% over the next five years in a report released in 2012. Lucintel highlights that the industry is likely to be driven by an increasing demand in Asia Pacific (APAC) and Europe, due to rising GDP, greater product awareness and a rise in organised retail. Although industry growth has been affected by recessions over the decades, it has managed to persevere through the economic downturns relatively unscathed. According to fellow market researcher Mintel, it was observed that despite poor sales in 2009 during the recent recession (2007–2009), the total value of the cosmetics and personal care sectors actually increased by 50% over the last decade in the UK market.[25] A phenomenon, dubbed as the "lipstick effect," was discovered back when researchers noted that the sales of personal care products actually rose when there was a recession. The underlying reasons for this have been contested for many years, ranging from an increase in "pampering" behaviour due to economic and personal unease to a desire to look your best when employment is high.[26]

4.1.2 Natural "Progression" of Personal Care Market

Market analysts predict that as consumers become increasingly aware of potentially toxic chemical ingredients used in cosmetics, market segments offering fragrance-free products made using natural ingredients and essential oils are likely to record strong gains.[24] For many years, cosmetic chemists were cynical about natural cosmetics and the effectiveness of natural cosmetic ingredients in comparison with their synthetic counterparts. However, the years following the zeitgeist of the 1960s, such as Rachel Carson's *Silent Spring*, along with environmental rights movement, helped to fuel a rising demand for effective products that were healthier for people and better for the environment.[26] Based on reports from market research firms, Transparency Market Research and Grand View Research, the global demand for organic personal care products is expected to grow annually at rates between 9–10% in the next five years.[27,28] Additionally, in the most recent recession (2007–2009), conventional personal care products took a substantial hit with industry giants such as L'Oreal whereas the natural personal care industry dropped from growth rates in the high teens to around 8–10% annually. Furthermore, the highly publicized acquisition of organic personal care industry leader Burt's Bees by Clorox as well as Colgate-Palmolive's purchase of Tom's of Maine within the last decade marked the recognition by market players of the health and wellness trend as a long-term direction that will shape the future of the business. As seen in Table 4.1, the growth rates indicate rising revenue and penetration of natural and organic personal care products into conventional market segments.[26] Hence, it is clear that a multitude of opportunities for growth exists in this maturing market.

4.2 Personal Care Formulations

Personal care products are mostly colloids, a mixture where one substance is evenly dispersed in another.[29] The particle size of the dispersed phase is generally between 5–200 nm in diameter so that it is too large to dissolve and too small to settle out easily like suspensions. Colloids employed in personal care can be found typically in the form of emulsions such as creams and lotions, liquid aerosols like hairsprays, foams like shaving foams or even gels. An emulsion can be defined simply as a heterogeneous system of two immiscible fluids in which one liquid is dispersed in another in the form of droplets.[30] The non-dispersed liquid is also known as the continuous phase. Depending on the continuous phase used, there are generally two types of emulsions: oil-in-water (o/w) or water-in-oil (w/o) emulsions (Figure 4.1).[3] Since most personal care products are emulsions, they usually consist of a basic combination of ingredients such as solvent (*e.g.* water), emulsifiers, preservatives, rheology modifiers/thickeners, colouring agents, fragrances and pH stabilisers.[2]

Table 4.1 Natural and organic personal care channel breakdown[26]

Product category	Total conversion sales	NOG sales/ penetration	NOG growth (%)
Cosmetics	$15.1 billion	$419 billion/2.8%	10.1
Female hygiene	$2.76 billion	$100 million/3.6%	15.1
Hair care/colour	$10.01 billion	$1.72 billion/17%	11.3
Baby care	$623 million	$159 million/25.5%	18.4
Nail care	$730 million	$24 million/3.2%	9.2
Oral hygiene	$4.91 billion	$605 million/12.3%	7.3
Bath items	$659 million	$147 million/22.2%	6.3
Deodorants	$1.61 billion	$217 million/13.5%	9.5
Shaving	$1.81 billion	$138 million/7.6%	8.7
Skincare	$9.20 billion	$3.05 billion/33.1%	6.6
Soap	$3.53 billion	$998 billion/28.2%	7.8
Fragrances/aromas	$5.91 billion	$352 million/6%	6.0
Total personal care	$56.98 billion	$7.92 billion/13.9%	8.4

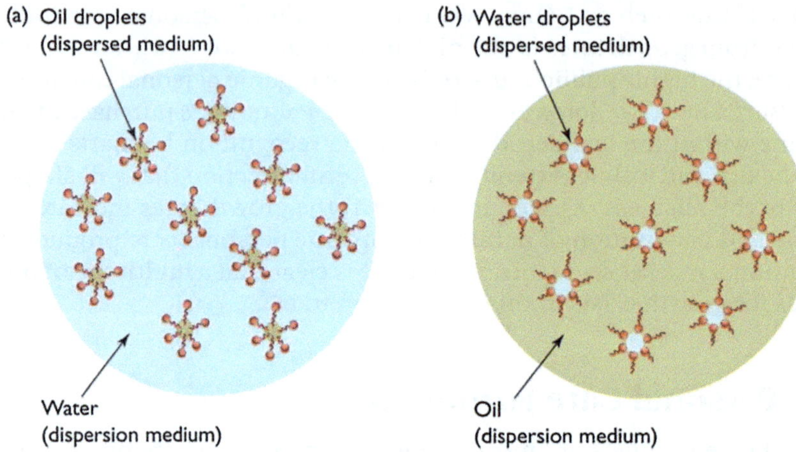

Figure 4.1 (a) Oil-in-water (o/w) emulsion, (b) water-in-oil (w/o) emulsion.[31]

4.2.1 Emulsifiers

Since many personal care products are based on emulsions containing oil and water, which are immiscible in each other, emulsifiers are added into the formulation to produce small droplets and prevent the oil and water phases from separating.[2] Hence, they are an important stabiliser for emulsion-based products. Most emulsifiers are considered surfactants as they function by reducing the surface tension between water and oil, thus producing a homogeneous product with an even texture. The surface-active properties of an emulsifier are related to its hydrophile–lipophile balance (HLB), which gives it an affinity for both water and oil. The HLB system determines the relative polarity of materials by ranking the most polar, water soluble materials at the top of the twenty-point scale while non-polar, oil soluble materials fall closer to zero. In an emulsion, the emulsifier's polar, hydrophilic groups (schematically depicted as a head) orient toward the polar water phase while their non-polar, lipophilic groups (schematically depicted as a tail) orient toward the oil phase to form micelles (Figure 4.1). The lipophilic tails of an emulsifier are usually composed of C16 fatty acid (palmitic acid) or longer fatty acids such as C18 fatty acid (stearic acid). The spherical structures of micelles provide stability to the dispersed droplets found in an emulsion through hydrogen bonding or weak electrostatic forces.[3]

Emulsifiers can be classified based on their ionic charge into cationic, anionic, amphoteric or non-ionic. Ionic emulsifiers dissociate into cationic or anionic species when solvated, such as sodium stearate or quaternium-24. Non-ionic emulsifiers are often used in skincare for their safe history of use and low reactivity. A common type of non-ionic cosmetic emulsifier is glyceryl esters such as glyceryl monostearate (GMS) and glyceryl distearate (GDS), which are prepared by direct esterification of glycerine. GMS and GDS are very effective emulsifiers as they contain both polar hydroxyl (OH) groups and non-polar fatty acids. Glycerol tristearate (GTS), on the other hand, does not contain hydroxyl (OH) groups so it possesses little emulsifying capability. Alternatively, sorbitan esters and their ethoxylates are a class of commercially available non-ionic emulsifiers under the brand names Span and Tween. Spans are sorbitan esters produced by the esterification of sorbitol and fatty acids. The HLB of Spans decreases with an increasing degree of esterification, conferring excellent solubility in lipophilic systems. Tweens are ethoxylated Spans that are hydrophilic in nature. The solubility of Tweens can be tuned by factors such as the degree of ethoxylation, number of ester groups or molecular weight of the fatty acid.[32]

4.2.2 Preservatives

Personal care formulations often contain preservatives so as to prevent microbial growth, which can reduce the product shelf-life as well as potentially harm the user. Parabens and ester of parabenzoic acid are the most commonly used as anti-microbial agents. Antioxidants such as tocopherol (vitamin E) and butylated hydroxytoluene (BHT) are sometimes included as well to prevent oxidation of sensitive ingredients. A mixture of preservatives is typically used to protect against various strains of bacteria, yeasts and moulds.[2,3]

4.2.3 Colouring Agents, Fragrances and pH Stabilisers

These ingredients are usually included in personal care formulations to impart aesthetic appeal by visual colour or smell. Colouring agents are either pigments or dyes that are especially essential in makeup cosmetics or hair colourings. pH stabilisers are particularly important for creating stable emulsion formulations as they help to maintain a balanced pH thereby avoiding excessive acidity or alkalinity.[2,3]

4.2.4 Rheological Modifiers

Rheology behaviour of personal care products is one of the most important features, not only from the technical but also from the aesthetic point of view. This can be achieved by using rheological modifiers, otherwise commonly known as thickeners, which are additives that are primarily used to increase the viscosity or impart a desired rheological profile to a formulation. However, they can sometimes be multifunctional and perform secondary roles such as gelling agents, emulsifiers, conditioners or film-formers.[33] In the area of personal care products, rheological modifiers improve the overall body, smoothness and silkiness, enhancing the aesthetic appeal.[34] These products are expected to be easy to use during application but also meet the sensory texture properties that are appealing to consumers. Consumers often tend to equate the viscosity of the product with quality. Hence, the viscosity of the

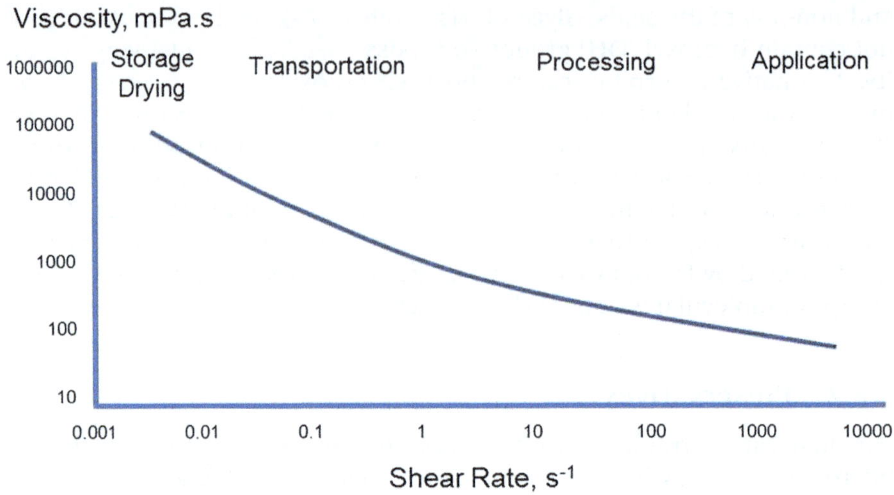

Figure 4.2 Effect of shear rate on application properties.

finished formulation should not be too thin or too thick. Furthermore, the required viscosity may vary at different stages of the product lifecycle from its manufacturing and processing to end use. Figure 4.2 shows an example of the required viscosities under different shear rate conditions. Generally, during processing when mixing, stirring and pumping are involved, low viscosities are essential but storage should take place at high viscosity so that settling out is prevented.[35] Therefore, the formulator has to decide what flow characteristics are needed across the entire spectrum of shear rates to which the product will be subjected.

Polymers represent a major class of ingredients that are used as rheological modifiers in personal care. This chapter aims to highlight the different types of natural rheological modifiers available for personal care on the market, recommended applications and their limitations. However, due to the exhaustive list of rheological modifiers commercially available, more emphasis will be placed on selected aqueous-based thickeners to meet the requirements of the Clean Air Act, which regulates the usage of volatile-organic compounds (VOCs) in cosmetic formulations.[1,36]

4.3 Rheology

Rheology, originating from the Greek words *rhéō* ("flow") and *–logia* ("study of"), is the study of deformation and flow of matter.[35] Flow is the continuous deformation of a material under the influence of external forces. When a force is applied to a liquid, it will flow to relieve the strain from this force. Different systems resist this flow more or less than others and the measurement of this resistance gives us the most frequently used rheological parameter, viscosity. The flow measurement of a liquid can be explained in terms of the

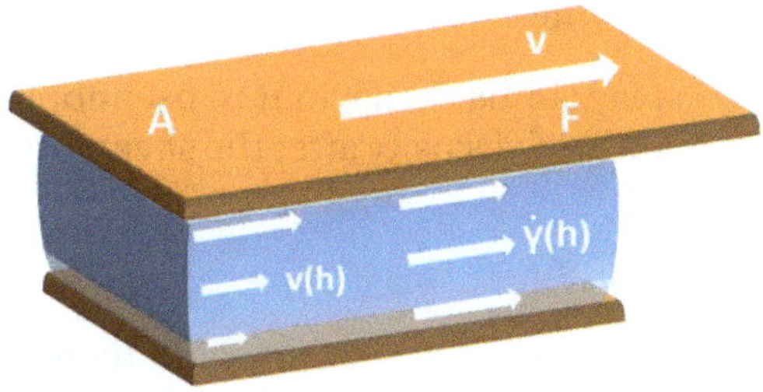

Figure 4.3 Simple Newtonian shear model.[37]

parallel-plate model introduced by Sir Isaac Newton where a fluid is flowing between two parallel plates, one moving at constant speed while the other one is stationary (Figure 4.3).[37]

Two characteristic quantities, shear stress (τ) and shear rate ($\dot{\gamma}$), can be derived from the model. Shear stress (τ), is the force (F) applied to the rectangular surface (A) when it is deformed by shear strain. The shear rate ($\dot{\gamma}$) of the flowing fluid is defined by the velocity v and the displacement h.[35,37]

$$\text{Shear stress}(\tau) = \frac{F}{A}\left[\frac{N}{m^2} = Pa\right] \qquad \text{Shear rate}(\dot{\gamma}) = \frac{v}{h}[\text{s}^{-1}]$$

Viscosity can be calculated from the shear rate and the shear stress according to the following equation:

$$\text{Viscosity }(\eta) = \frac{\tau}{\dot{\gamma}}[Pa \cdot s]$$

4.3.1 Rheology Profiles

Rheological modifiers may exhibit different rheological behaviours which are useful for various kinds of applications in personal care.

4.3.1.1 Newtonian Flow

If the viscosity of a substance is constant at different shear rates, it is said to exhibit ideal or Newtonian behaviour. Newtonian flow is generally observed in only low molecular weight liquids such as water, solvents and mineral oils.[35,37,38]

4.3.1.2 Pseudoplastic Flow

In reality, most polymer solutions show a flow behaviour, where viscosity decreases with increasing shear rate. However, when the shear force is removed, the fluid immediately reverts back to its original viscosity. This is known as shear thinning or pseudoplastic behaviour.[35,37,38]

4.3.1.3 *Thixotropic Flow*

Thixotropic liquids display similar behaviour to pseudoplastic flow but there is a time-dependent rebuild of viscosity after the shear force is removed.[35,37,38]

4.3.1.4 *Dilatant Flow*

Materials that exhibit an increase in viscosity with increased shear rates are dilatant or shear thickening. This can be found in dispersions with high solids content or high polymer concentrations.[37,38]

4.3.2 Types of Rheological Modifiers

Rheological modifiers can be classified by a variety of schemes, including their ionic charge (anionic, cationic, non-ionic or amphoteric), their application in aqueous or solvent-based formulations and their thickening mechanisms. Traditionally, rheological modifiers can be classified by their chemical nature (inorganic and organic) and their origin, whether they are natural, naturally-derived or synthetic. Figure 4.4 provides a general overview and categorization of the major rheology technologies used in personal care.

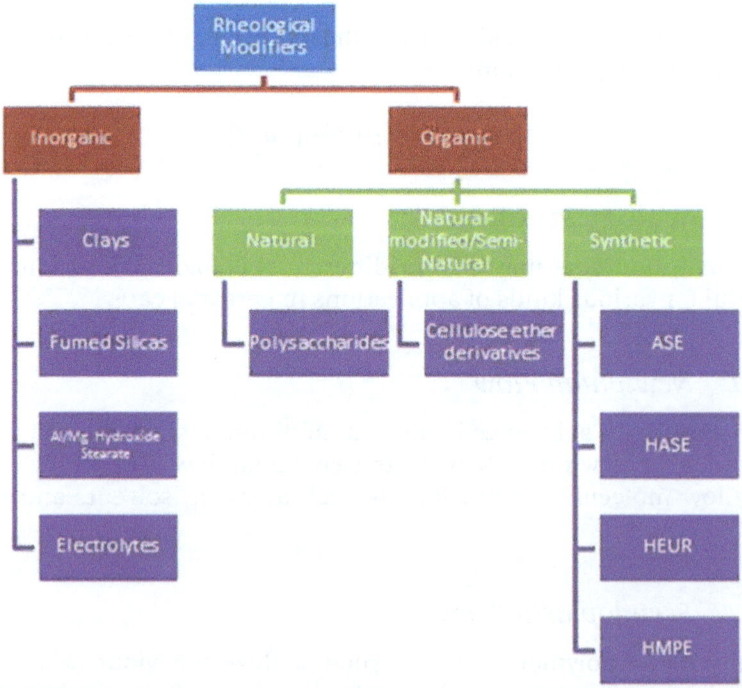

Figure 4.4 Overview and classification of major rheology modifiers for personal care.

4.3.2.1 Natural Polymers: Polysaccharides

Natural polymers, completely natural or semi-natural, are usually derived from plant, animal or microbial origins so they are large polymers with chemical structures based on proteins or polysaccharides. Polysaccharides are a class of natural polymers which has been ubiquitously used in cosmetics. Historically, they have been employed in personal care formulations due to their ready availability from common natural sources and their unique multifunctionality. The interplay of polysaccharides with other formulation ingredients and their facile chemical modification could enhance the properties of personal care products. Lastly, these naturally occurring polysaccharides are renewable and have a history of safe usage as opposed to synthetic-based polymers. Polysaccharides are capable of performing a myriad of cosmetic functions such as rheological modifiers, suspending agents, moisturisers, emulsifiers, emollients, hair conditioners or even wound-healing agents. The multifaceted functionality of polysaccharides poses a challenge for anyone trying to understand a single rheological influence of a polysaccharide in a formulation.[39] Polysaccharides are typically manufactured within living systems to perform natural functions that may differ significantly from that in a personal care formulation. As an example, starch is commonly manufactured in potatoes as food or energy storage but not as a rheological modifier or gelling agent, for which it is used in personal care. The manufacture of polysaccharides in living biological systems is tightly controlled by the organism's genes and enzymes, thus resulting in little molecular variation but also giving rise to polymeric consistency.[39]

Polysaccharides, sometimes referred to as "polyglycans," can be best thought of as long molecular chains made up of building block units known as monosaccharides bonded together by an essential anhydroglycosidic bond. The structure of these chains can be linear (*e.g.* cellulose) or branched (*e.g.* guar gum) depending on the type of anhydroglycosidic bond that is formed between monosaccharides. The presence of branching in a polysaccharide can greatly influence its solution behaviour as it tends to disrupt interpolysaccharide bonds that would stack the polysaccharide chains close to one another. Likewise, highly branched polyglycans occupy greater solution volume when dissolved than linear ones. Highly branched polyglycan chains also entangle more easily, increasing the solution viscosity. Also, highly branched polyglycans usually dissolve more rapidly due to the disruption of crystallinity by branching.[39]

Many polysaccharides form helical structures, analogous to that of proteins. This three-dimensional spatial configuration is arranged to minimise the total energy of the polysaccharide. The nature of the helix may vary from a tight, spring-like coil (*e.g.* amylose) to a loose, open helix (*e.g.* cellulose), twisting left or right. Like polypeptides, they can further minimise their internal molecular energy by wrapping one helical coil around another, often assuming the double-helix conformation. The double-helix conformation can be further influenced by supramolecular bonds such as interpolysaccharide hydrogen bonding, van der Waals forces or steric influences and positioning of groups between polysaccharide chains. Strong attractive influences can create highly ordered or crystalline segments but when weaker or repulsive form amorphous regions and even cause chain separation. These effects are closely tied to the understanding of the polysaccharide's solution behaviour. However, generally regardless of the structure in solution, its effect on viscosity becomes significant as the concentration increases and individual chains begin to interact with each other. Independent of the polysaccharide, it is observed that the concentration must exceed a critical value, known as the critical overlap concentration, C^*, before interaction of helices or coils occurs.[40] This can be graphically represented in plots of specific viscosity, η_{sp}, *versus* the coil overlap parameter, $c[\eta]$ (Figure 4.5). As seen in Figure 4.5, at C^* where $c[\eta] \approx 4$, the shape of the viscosity curve changes and the viscosity in this concentrated regime is controlled by chain entanglement.[39,41]

The solution behaviour of polysaccharides also depends heavily on the nature of the substituent groups bound on the individual monosaccharides, be it of natural origin or synthetically modified. Regardless of their origin, polysaccharides can be broadly divided based on their ionic charge: anionic, cationic, non-ionic, amphoteric (Figure 4.6).[39,42]

4.3.2.2 *Anionic Polysaccharides*

This class of polysaccharides, which possess a negative charge, is predominated by an extensive list of naturally occurring materials. Commercially, there are only two derivatives that are of rheological interest to personal care: cellulose gum (carboxymethylcellulose) and carboxymethylchitin. These derivatives are modified to be anionic by the carboxymethylation of naturally occurring cellulose and chitin respectively.

4.3.2.2.1 Alginates (Alginic Acid). Alginates are isolated from marine brown algae and from bacterial fermentation and are comprised of two different monosaccharides, β-D-(1,4)-mannuronic acid and α-L-(1,4)-guluronic acid (Figure 4.7).[39,43–47] These two monosaccharides can be linked glycosidically as polymannuronic acid, polyguluronic acid or as combinations. The β-(1,4)-linkage of D-mannuronic acid gives alginates an extended helical structure while the α-(1,4)-linkage of L-guluronic acid gives it a tighter helical coil. Therefore, alginates comprising primarily D-mannuronic acid residues tend to be extended coils in solution while the presence of L-guluronic acid compresses the coils.

Both monosaccharides, β-D-(1,4)-mannuronic acid and α-L-(1,4)-guluronic acid, have carboxylic acid groups, making them highly anionic and influencing the solution behaviour of alginates. In the presence of polyvalent cations (*e.g.* Ca^{2+}), cross-linking sites can occur (called "junction zones") due to strong intra- and intermolecular bonding with polysaccharides containing L-guluronic acid residues.[39,48–52] The relationship between the alginate (the

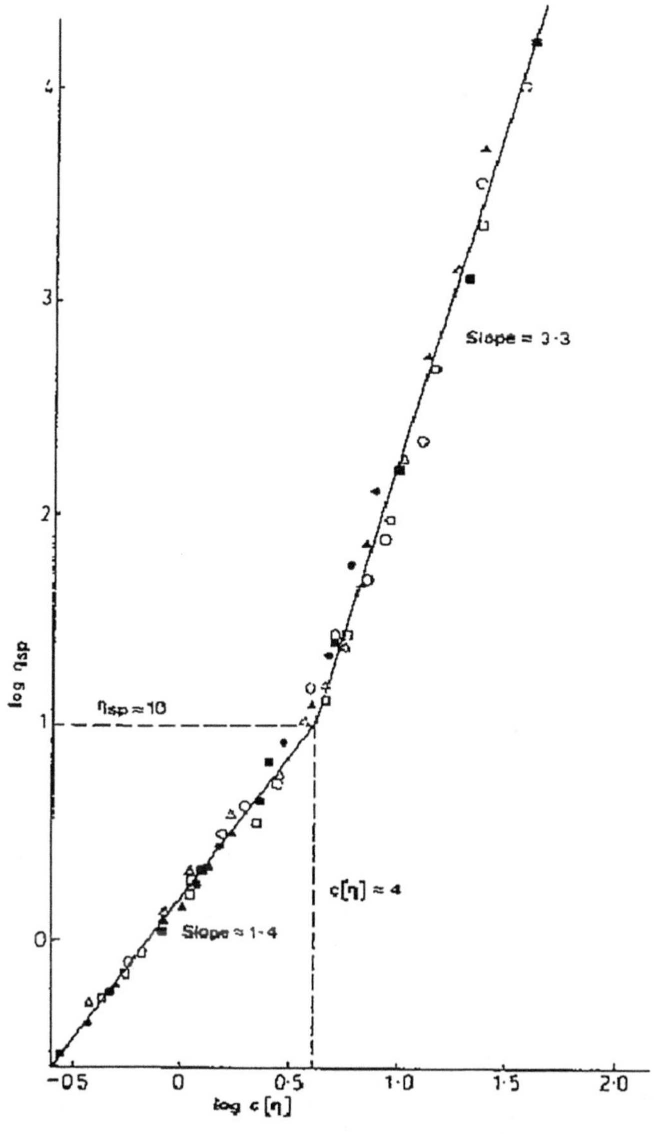

Figure 4.5 Plot of $\log \eta_{sp}$ *vs.* $\log c[\eta]$ for various polysaccharides.[39,41] Reprinted from E. R. Morris *et al.*, Concentration and shear rate dependence of viscosity in random coil polysaccharide solutions, *Carbohydr. Polym.*, 1981, **1**, 5–21. Copyright (1981), with permission from Elsevier.

"box") and the cation (the "egg") can be described using an "egg box" model, as seen in Figure 4.8. The viscosity of aqueous alginate solutions increases with the concentration of polyvalent cation until a gel is eventually formed. Further addition of Ca^{2+} ions will however cause precipitation of the alginate from the solution. Therefore, the polyvalent metal cation concentration

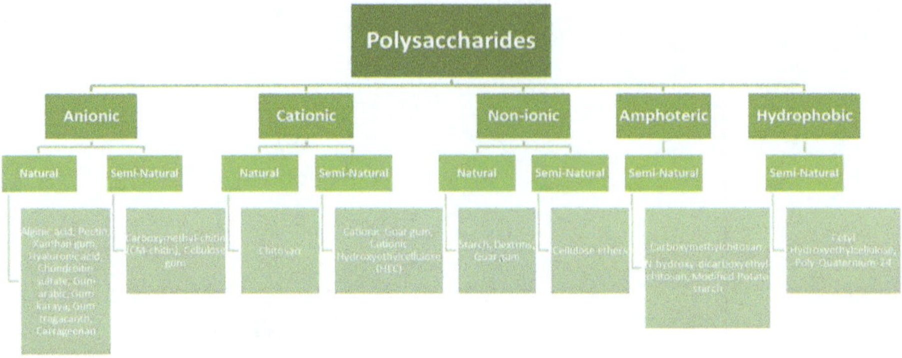

Figure 4.6 Classification of polysaccharides used in personal care.[42]

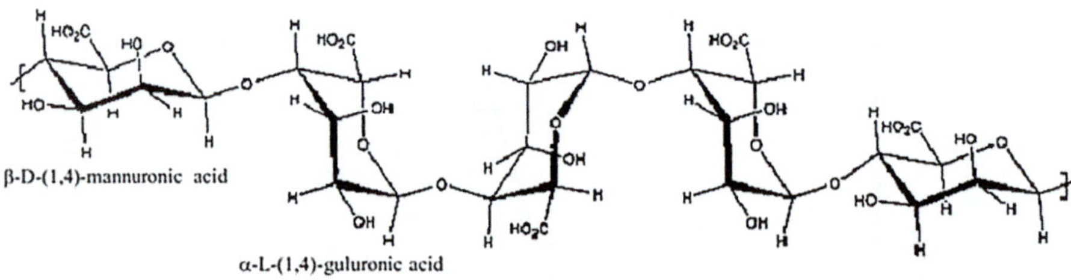

Figure 4.7 Partial structure of alginic acid.[39]

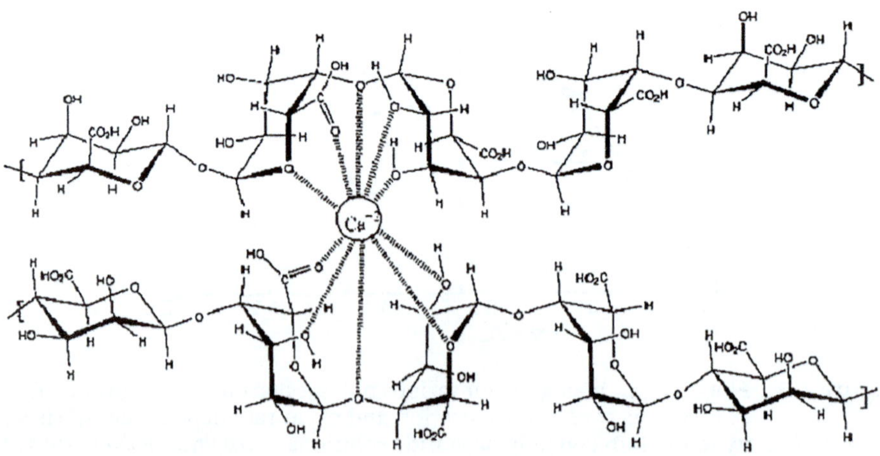

Figure 4.8 "Egg box" model illustrating junction zones formed between alginic acid and Ca^{2+} ions.[39]

(Ca^{2+} ions can be found in tap water) must be carefully controlled through the use of sequestrants such as ethylenediaminetetraacetic acid (EDTA) to optimise the thickening effects of alginate.[39]

Like most anionic polysaccharides, aqueous alginate solutions are influenced by pH due to the protonable carboxylic acid groups. The effect of pH

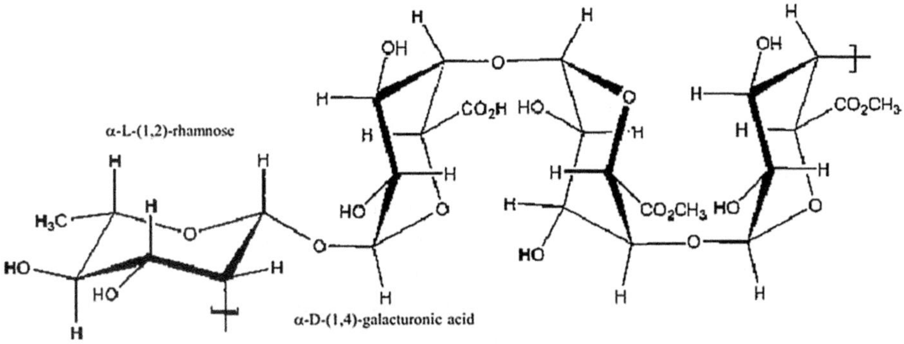

Figure 4.9 Partial structure of pectin.

is most pronounced near the isoelectric point of carboxylic acid near pH 4.0, where below this pH, solution viscosity increases sharply due to protonation of carboxylic acid salts (Figure 4.9).[39,43] The recommended pH for use is above 3.0 as alginates are unstable at low pH owing to hydrolysis of glycosidic bonds below pH 3.0.[39]

4.3.2.2.2 Pectin. Pectin is isolated industrially by extraction from citrus fruit peels and shares a structure similar to alginic acid. It is comprised of repeating α-D-(1,4)-galacturonic acid units, which are interrupted occasionally by α-L-(1,2)-rhamnose, a form of mannose with a fully reduced methyl group (Figure 4.9).[39,43,45,46,53] Each C6 carbon atom can be fully oxidised to a carboxylic acid group imparting a negative charge but naturally, some of the carboxylic acid groups are esterified with O-methyl groups. This renders pectin partially non-ionic and results in gels that behave differently from those of non-esterified alginates.[51] As such, pectin is usually sold in grades designated by the level of methyl ester, with lower levels making pectin more anionic. At methyl ester levels below 50%, pectin behaves like alginic acid, forming turbid gels in the presence of divalent metal ions (Ca^{2+}). At high levels, it is still capable of forming gels without divalent ions but the concentration of pectin required is higher. Similarly, pectin functions well at pHs near 3.5 and exhibits unstable viscosities at the pH extremes where hydrolysis is rapid.[39]

4.3.2.2.3 Xanthan Gum. Xanthan gum is a polysaccharide that is commercially produced by precipitation of the broth isolated from the exocellular coating on bacterial cell walls after fermentation. It has a complex structure comprising of a backbone of β-D-(1,4)-glucose with branching trisaccharide side chains of β-D-(1,2)-mannose and β-D-(1,4)-glucuronic acid. The comb-like trisaccharide side chains found on the primary chain confers it water solubility. The anionic charge of xanthan gum is by virtue of the carboxylic acid residues on D-glucuronic acid and the pyruvic acid moiety on the terminal D-mannose.[39]

It is generally accepted that xanthan gum forms helical coils in the presence of cationic salts but whether double helices are formed is still debatable.[51] The thickening and gelation mechanism requires formation of junction zones, hypothesized to occur through hydrophobic interactions on the helix. Cations tend to "salt out" hydrophobic sites and improve the strength of the junction zones, producing a slight increase in viscosity although the need for salts to build solution viscosity is minimized by the rigid backbone. The viscosities of xanthan solutions also show excellent thermal stability as their viscosity is derived from hydrophobic interactions. Xanthan solutions are pseudoplastic with a yield value, below which the solution resists deformation. Such solutions are suitable for stabilising or suspending water insoluble particles or oily droplets. This prolongs the lifetime of formulations and might reduce the amount of primary emulsifiers needed. Xanthan gum forms cholesteric mesophase lyotropic liquid crystals in water at concentrations as low as 3.5 wt%, an area of expanding interest in personal care due to their unique ability to form emulsifier-free suspensions. Formation of liquid crystals is influenced by the salt concentration, with higher salt concentration requiring lower xanthan concentration. Xanthan gum exhibits rheological synergy with gluco- or galactomannans (*e.g.* kappa-carrageenan) as the number and strength of junction zones are likely to increase causing higher solution viscosities.[39]

In terms of compatibility with surfactants, anionic xanthan gum is incompatible with most cationic surfactants. Independent of the surfactant concentration, the viscosities of xanthan solutions are relatively unaffected by non-ionic surfactants.[54] The presence of anionic surfactants at concentrations greater than 4 wt% increases the viscosity of xanthan solutions, especially at low shear rates. This further improves the suspending capability of xanthan gum. In addition, the thickening efficiency of xanthan gum can be synergistically enhanced by addition of colloidal magnesium silicate (Veegum).[39]

4.3.2.2.4 Carrageenans. Carrageenans are anionic polysaccharides derived from red algae (*Rhodophyceae*) with at least seven varieties being known.[38,43,45,46,55,56] The most commercially available varieties, kappa, iota and lambda, primarily comprise of two repeating monosaccharides, β-D-(1,3)-galactose and α-D-(1,4)-galactose (Table 4.2), each differing in the number and position of the sulfate ester groups.[57] The anionic character of carrageenans is due to the substitution of some of the pyranose hydroxyl groups. The presence of sulfate ester groups increases the acidity and hydrophilicity of the carrageenan. As such, in aqueous solutions, kappa- and iota-carrageenans form double helices peppered on the exterior with hydrophilic anionic sulfate groups.[39]

Temperature is required to provide stimulus for the formation of kappa- and iota-carrageenan double helices. Carrageenans exist as random coils when they are initially dissolved in hot water. Upon cooling, the hydrophobic domains of kappa- and iota-carrageenans interact unfavourably with water, resulting in the aggregation of the chains by hydrophobic interactions, forming a single helix first then eventually a double helix (Gel I) (Figure 4.10).[43] Additionally, the rheological properties of kappa-carrageenans are dependent on molecular weight as well. The greater the molecular weight of the polysaccharide, the higher its aqueous solution viscosity.[58] In contrast, lambda-carrageenans do not even form single helices and are unable to gel as a result of their higher hydrophilicity.[38] Even so, they are still able to thicken aqueous solutions by random chain entanglements.[39]

In the presence of cations, especially K^+ and Ca^{2+}, the double helices can further aggregate by formation of junction zones through the anionic sulfate groups.[51,59–62] Cationic salts are important for the formation of interhelical

Table 4.2 Structure of kappa-, iota- and lambda-carrageenans[39]

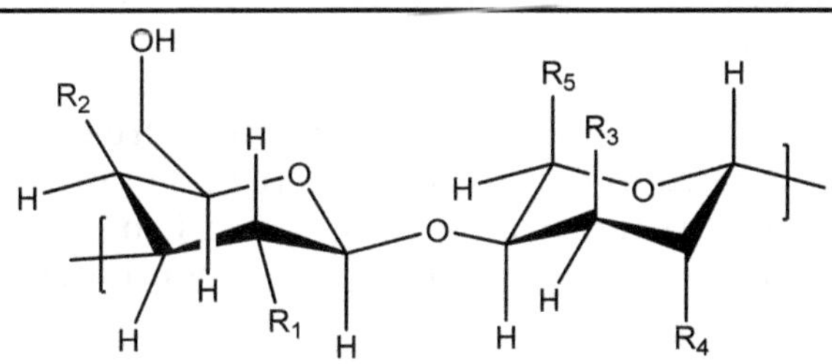

β-D-(1,3)-Galactose | α-D-(1,4)-Galactose

Carrageenan type	R_1	R_2	R_3	R_4	R_5
Kappa	OH	OSO_3^-	O (linked to R_5)	OH	CH_3
Iota	OH	OSO_3^-	O (linked to R_5)	OSO_3^-	CH_3
Lambda	OSO_3^-	OH	OH	OSO_3^-	$CH_2OSO_3^-$

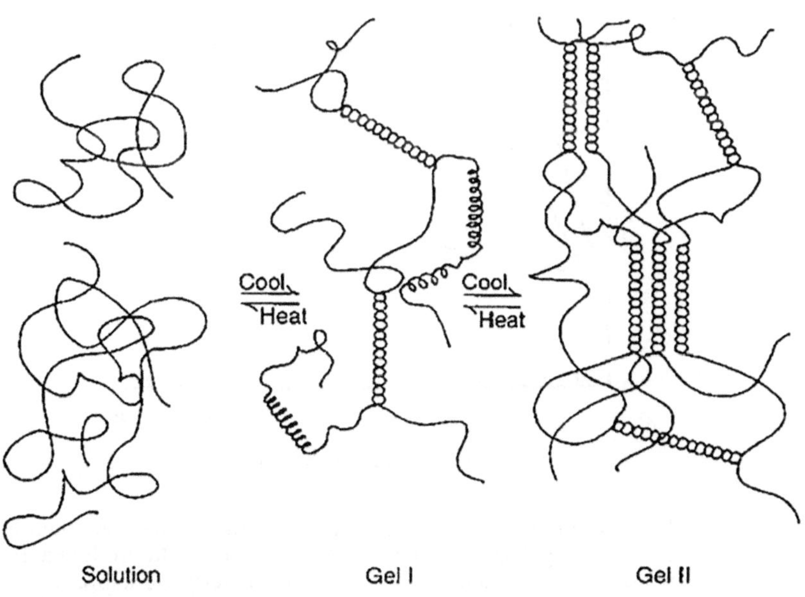

Figure 4.10 Coil to aggregated helix transitions for gelling carrageenans.[39,43]

cross-links through junction zones that increase the solution viscosities as well as forming strong gels (Gel II).[39,43] However, carrageenans are unstable at pH values below 3.5 as sulfate ester groups are prone to acid hydrolysis, which accelerates at high temperatures.[38]

Kappa-carrageenans also display an interesting synergistic thickening effect in blends with galactomannans or glucomannans (*e.g.* locust bean gum). This can be seen from the manifestation of higher solution viscosities and elastic moduli, as seen in Figure 4.11. The exact mechanism is still being debated while the prevailing theory attributes the effect to the binding of gluco- and galactomannans to the carrageenan double helix, disrupting the salt-induced aggregation of double helices, increasing the number and strength of junction zones.

4.3.2.2.5 Cellulose Gum (Sodium Carboxymethylcellulose). Cellulose gum, otherwise termed sodium carboxymethylcellulose (Na-CMC), is a chemically modified derivative of cellulose, which is naturally found in all plants.[43,46,64] It is commercially isolated from cotton linters or wood after

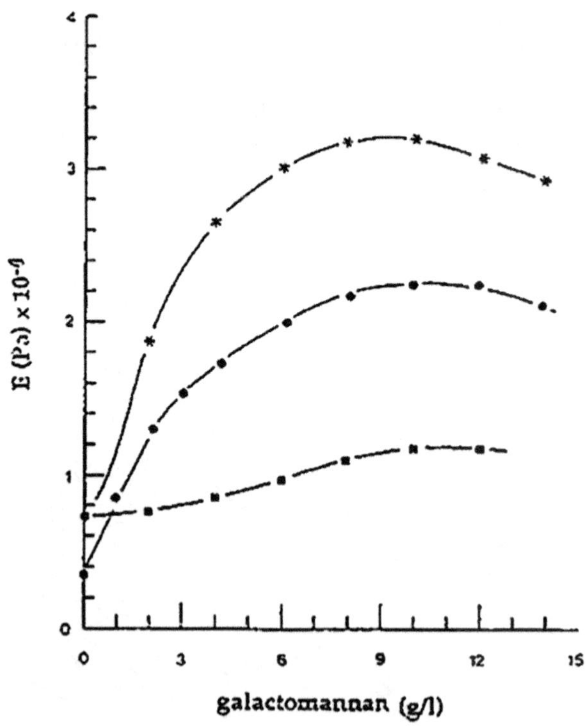

Figure 4.11 Elastic modulus of kappa-carrageenan–galactomannan blends as a function of galactomannan concentration.[39,63] Reproduced from T. Turquois *et al.*, Rheological studies of synergistic kappa caarageenan-carob galactomannan gels, *Carbohydr. Polym.*, 1992, **4**, 263–268. Copyright (1992) with permission from Elsevier.

removal of other components such as lignin and hemicellulose. Naturally occuring cellulose is composed of one repeating monosaccharide, β-D-(1,4)-glucose (Figure 4.12). However, due to the strong intermolecular hydrogen bonding, cellulose is highly crystalline and ordered, making it water insoluble. One of the ways to counter this problem is the derivatization of cellulose with sodium chloroacetate to produce Na-CMC. Cellulose fibers are swelled with aqueous caustic and reacted with sodium chloroacetate in an aqueous organic diluent, in which both cellulose and CMC are swellable but insoluble. The degree of substitution (DS) is the average number of carboxymethyl groups attached to each glucose monosaccharide. Commercially available Na-CMC typically has a DS between 0.7 and 1.2.[39,57] The presence of substituted carboxymethyl groups makes Na-CMC highly anionic with an acidity similar to acetic acid. At a pH above 4.5, most of the acid groups are deprotonated in the salt form, hence Na-CMC is water soluble. CMC is water insoluble at a pH below 4.5, with viscosity first increasing as chain entanglement increases and then with phase separation. When the phase separation is completed, the solution becomes hazy due to precipitation, which could pose problems for clear formulations.[39]

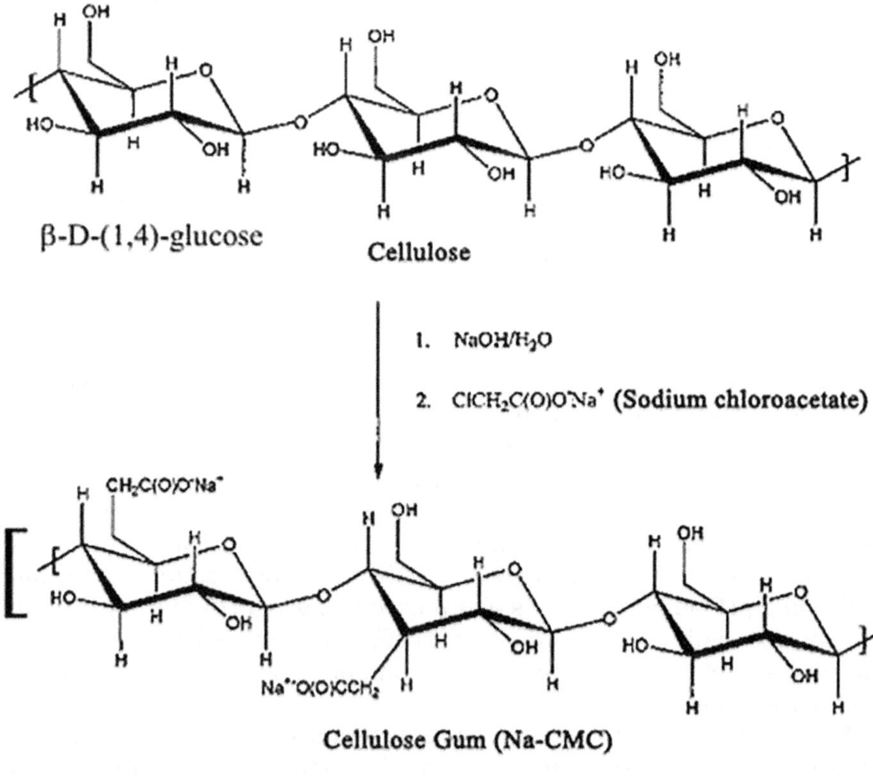

Figure 4.12 Partial structure and synthesis of cellulose gum.[39]

CMC solutions at concentrations as high as 5 wt% are pseudoplastic but do not possess a yield value. CMC is capable of dissolving in and thickening glycerin solutions with a small amount of water added for improved solubility. Like most anionic polysaccharides, CMC binds strongly to certain multivalent cations such as Ca^{2+}, causing it to gel. The compatibility of CMC with cations depends ultimately on the ability of the added cation to form a soluble salt of CMC. For example, potassium ions when added have little effect on solution viscosity and clarity, while zirconium ions result in precipitation when added. A general rule of thumb is that monovalent cations form soluble salts of CMC, divalent cations are borderline and trivalent cations cause precipitation of insoluble salts. However, there are exceptions to this rule as the effect of the cation varies with each particular cation, its concentration, pH, DS of CMC and the order of addition of CMC and cation. Increased tolerance for cations can be obtained with highly substituted CMC (*i.e.* DS = 0.9–1.2) or by dissolving the CMC before adding the salt.[65] CMC's thickening mechanism relies on the entanglement of its high molecular weight chains and not on the presence of divalent metal ions for formation of cross-linked junction zones. However, there are studies that suggest that partially substituted CMC forms more viscous aqueous solutions than fully substituted CMC due to the formation of interpolysaccharide junction zones through hydrogen bonding of pyranose rings in the modified crystalline regions.[39] Na-CMC has a pseudoplastic flow behaviour which tends to switch to thixotropic with increasing DS. However, its viscosifying effects are only stable within the pH range 4–10 and at lower temperatures.[38]

4.3.2.3 Cationic Polysaccharides

Cationic polysaccharides of interest to personal care consist mainly of synthetically modified polyglycans. Due to their cationic character, they tend to bind tightly to anionic surfaces such as human hair and skin. Hence, many cationic polysaccharides have found their use as conditioners or film-formers in cosmetic formulations for hair.[39]

4.3.2.3.1 Chitosan. Chitosan is well-known for being the only naturally occurring, commercially available polysaccharide to carry a cationic charge. It can be naturally found in insect exoskeletons and bacterial parasites but is industrially manufactured by alkaline hydrolysis of the acetamide of crustacean chitin.[66] Chitosan is a random copolymer made up of two monosaccharides, *N*-acetyl-β-D-(1,4)-glucosamine and β-D-(1,4)-glucosamine, generally in a 1 : 4 ratio commercially. However, chitosan is only cationic and water soluble at a pH below 7.0 as it is neutralised to a water soluble chitosonium salt. Commercially available chitosan is usually neutralised with a carboxylic acid such as lactic acid or glycolic acid, components of the natural moisturising factor found in human skin.[39]

Chitosan, in its salt form, has strong substantivity to negatively-charged hair and skin. It is also a film-former used in hair fixatives and shampoos. The strong cationic charge on the chitosan backbone causes incompatibility with most anionic surfactants and some polyvalent metal cations (*i.e.* Cu^{2+}) as it forms water insoluble complexes. Its biocompatibility with the human body has also seen its use in polysaccharide matrices for controlled drug delivery applications.[67] It has also been shown that chitosan is mucoadhesive, being able to bind to the glycoproteins of body tissue lining the eyes and throat.[39] This suggests possible future applications of chitosan in advanced cosmetics ("cosmeceuticals") that involve the delivery of pharmaceuticals.

4.3.2.3.2 Cationic Hydroxyethylcellulose (HEC). Cationic hydroxyethylcellulose (HEC) is a polysaccharide that found its most successful use as a conditioner in haircare formulations. As mentioned earlier, cellulose in its native form is water insoluble but it can be treated with reactive alkylating reagent, ethylene oxide, under alkaline conditions to form a water soluble cellulose derivative, HEC. HEC is extensively used in personal care as an aqueous-based rheological modifier. HEC can be further altered with various cationic reagents, which add quaternary charges randomly along the HEC backbone. Cationic HEC derivatives can be described by their cationic substitution (CS) level, the molar amount of quaternary nitrogen grafted onto the backbone. Two examples of commercially used cationic HEC derivatives, known commonly by their International Nomenclature Cosmetic Ingredient (INCI) names are polyquaternium-10 and polyquaternium-4 (Figure 4.13).

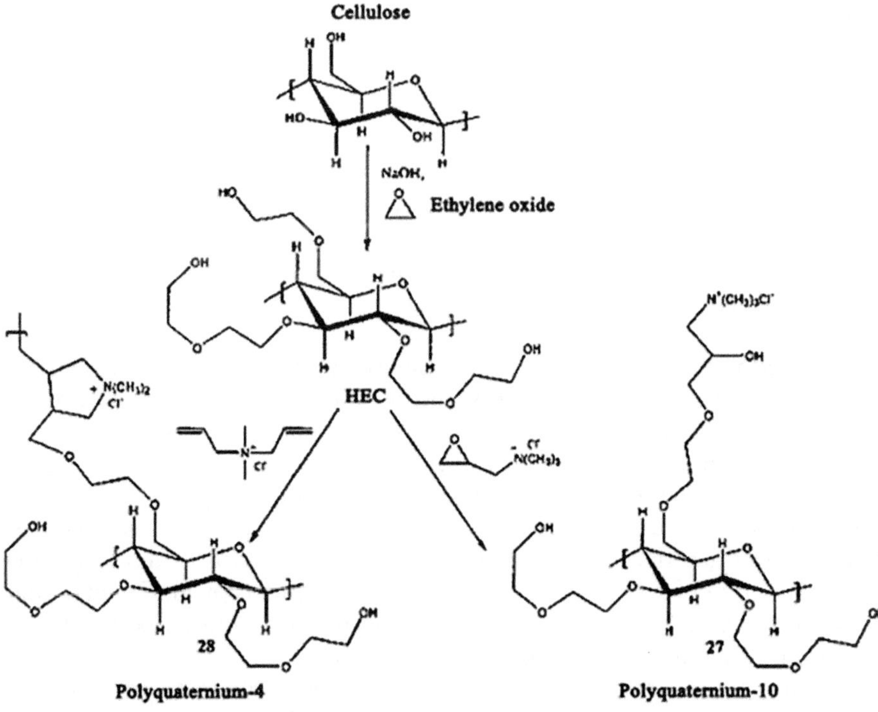

Figure 4.13 Synthesis and structures of HEC and its cationic derivatives.

Polyquaternium-10 is used as a conditioner in shampoo and conditioner formulations whereas polyquaternium-4 is used both as a conditioner and fixative in hair fixative formulations like mousses.[39]

It is generally accepted that cationic cellulose-based polysaccharides interact weakly with cationic surfactants/surfaces. In contrast, cationic polysaccharides like cationic HEC respond strongly to anionic surfactants/surfaces even at concentrations well below the surfactant's critical micelle concentration (cmc). The thickening efficiency depends on the molecular weight of cationic HEC; increasing molecular weight shows increased solution viscosity. The viscosifying effect is also concentration dependent.[68,69] In 1 wt% polyquaternium-10 solutions, a rapid increase in viscosity is observed as the anionic surfactant begins to neutralise the cationic charge on the polysaccharide. In comparison with Na-CMC mentioned previously, this behaviour is not observed with the anionic cellulose–cationic surfactant combination. In dilute 0.1 wt% polyquaternium-10 solutions, the presence of anionic sodium dodecyl sulfate (SDS) suppresses viscosity upon charge neutralisation. The charge neutralised polymer–surfactant complex is forced to phase separate by precipitation regardless of the polysaccharide concentration.[70] Depending on the surfactant type, the cationic HEC may redissolve or precipitate as surfactant concentration continues to increase. At high concentrations of surfactants, the rheological contributions of cationic HECs are greatly shadowed by the greater contributions of the surfactant.[39]

Similar to chitosan, polyquaternium-10 also has mucoadhesive properties, expanding its potential use in personal care as ocular or oral therapeutics.[39,71]

4.3.2.4 Non-Ionic Polysaccharides

Although non-ionic polysaccharides do not carry a formal charge, their solution behaviours can still be affected by changes in solvent polarity, presence of salts, surfactants, polymers and other components. Typically, they have found their usefulness in cosmetics as rheological modifiers, such as the family of ether modified cellulose derivatives.

4.3.2.4.1 Starch. Starch is one of the most abundant polysaccharides stored in all green plants for energy and is commercially isolated from different crop plants such as corn, potato, rice, wheat and tapioca. Industrial production involves the separation of the starch "granule" from other plant lipids, proteins and hemicellulose by some kind of milling process. The granule consists of two basic polysaccharides, amylopectin and amylose. Starches can be differentiated by their amylopectin to amylose ratio. Structurally, amylopectin is branched and can have molecular weights typically greater than 500 million g mol^{-1}. Amylose is linear and assumes a tight helix in solution with lower molecular weights of 1 million g mol^{-1}.

The main advantages of starch are its low economical cost and non-toxicity, being primarily used as a rheological modifier in personal care. The viscosities that can be achieved by starch range from low like water to high like that of gels. The viscosity can be correlated with the ratio of amylose to amylopectin, with higher ratios producing higher viscosities. Upon dissolution in hot water, amylose disperses readily unlike its diastereomer, cellulose. As the solution is cooled, it undergoes a process of retrogradation, in which it intermolecularly intertwines with itself to form firm gels. The intertwining is primarily driven by the formation of hydrogen bonds. Unfortunately, the strong hydrogen bonding reduces the ease of dissolving amylose in water, thus commercial grades frequently come in the form of a precooked powder. Visually, starch solutions also tend to be hazy, limiting their use in clear formulations. Solution clarity can be improved by removing the residual lipids and proteins but the necessary procedures are challenging and costly. Its thickening effects can be enhanced when used in combination with inorganic rheological modifiers such as bentonite.[39,72]

Starch is also known to form complexes when uncoiled (*i.e.* high temperature) which could reduce the effectiveness of surfactants or other key functional cosmetic ingredients such as fragrances and actives.[73]

4.3.2.4.2 Cellulose Ethers. Cellulose ether derivatives represent the most popular and useful rheological modifiers in personal care due to their history of safe and effective use. The available cellulose ethers that can be synthesized from cellulose are as shown in Figure 4.14.

Cellulose ether derivatives in general dissolve in water at room temperature by hydration of the cellulose backbone, developing a sheath of tightly bound water molecules around each chain. This allows the chains to expand, uncoil and dissolve to create the entanglement network that imparts viscosity. Significant viscosity can be achieved at low concentrations, mainly due to entanglement of high molecular weight, rigid cellulose molecules. However, cellulose ethers do not display yield values so they cannot act as a suspending agent without the aid of other materials such as xanthan gum.

The nature of the hydration sheath is affected by factors such as solvent polarity, presence of polymers or surfactants. Salts increase the solvent polarity, which reduces the solubility of more lipophilic cellulose ethers such as hydroxypropylcellulose (HPC) and hydroxypropylmethylcellulose (HPMC). Surfactants may increase the solubility or viscosity of HPC and HPMC. The solubility of most cellulose ethers decreases as the temperature is increased. Thermal energy breaks the hydrogen bonding between water molecules and the cellulose chain until the polysaccharide collapses and precipitates from the solution at a temperature known as the cloud point (cp).[39]

4.3.2.4.3 Hydroxyethylcellulose (HEC). HEC is the most hydrophilic and widely used cellulose ether. Unlike other cellulose ethers, it has a more or less non-existent cp of 100 °C at atmospheric pressure.[43,46,74] It is not soluble in many organic solvents used in cosmetics. Due to the lack of hydrophobic interactions of methyl groups, HEC is more tolerant to the effects of pH extremes, salts, surfactants. Hydration of HEC can be troublesome because

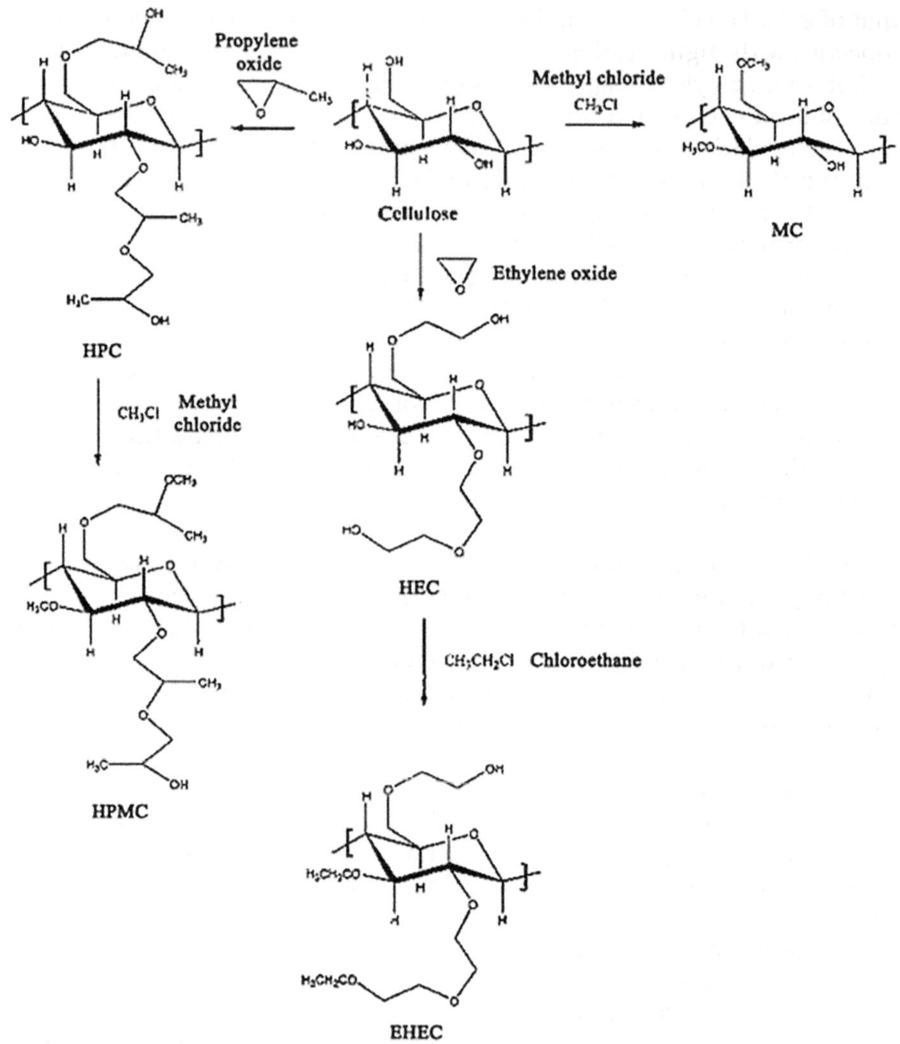

Figure 4.14 Synthesis routes for cellulose ether derivatives.[39]

of its cp. Extra caution should be exercised to ensure that partially hydrated spheres ("fisheyes") do not form during the dissolution of HEC.[39]

4.3.2.4.4 Hydroxypropylcellulose (HPC). HPC has a cp of about 45 °C.[43,46,74] Hydration of HPC is best done by adding to hot water to fully disperse and, upon cooling, it dissolves completely without clumping. HPC is more lipophilic than HEC, allowing it to dissolve and thicken many organic solvent-based systems. Similar to xanthan gum, the formation of hydrophobic domains by methyl groups on HPC allows for cholesteric lyotropic liquid crystals to be formed. In solutions of HPC and SDS surfactant, it is observed that the self-assembly of SDS is nucleated by lipophilic domains on HPC at concentrations well below the cmc. This may increase the effectiveness of surfactants in formulations by reducing the concentration of surfactants needed. Increasing salt concentrations causes hydrophobic interactions to cross-link polysaccharide chains, which is accompanied by an initial increase in viscosity. Eventually, HPC is salted out showing a viscosity reduction. HPC is also relatively insensitive to pH fluctuations as it is non-ionic.[39]

4.3.2.4.5 Methylcellulose (MC). Methylcellulose (MC) has a gel point around 45 °C and can be hydrated in the same fashion as for HPC.[43,46,75] At temperatures above 56 °C, it precipitates from the solution, creating a hazy, cloudy appearance. MC displays similar solution behaviour as HPC but tends to be more lipophilic. This allows it to dissolve and thicken a wider range of organic solvents. MC is compatible with most surfactants. MC has a unique ability to improve foam generation and strength in shower formulations. In hot/warm showers, MC approaches its gel point and increases the viscosity of the films surrounding bubbles, making bubbles and foam more durable.[39]

4.3.2.4.6 Hydroxypropylmethylcellulose (HPMC). HPMC is similar to MC except for the presence of hydroxypropyl groups which make up for the hydrophilic domains lacking in MC. Hence, HPMC can function better in some solvent systems than MC. Interaction between SDS surfactants and HPMC in solution shows that the surfactant micelles create more intermolecular chain entanglement than seen between SDS and HPC. The concentration of HPMC must exceed a critical overlap concentration, C^*, or the polysaccharide tends to collapse and viscosity decreases. At very high surfactant concentrations (>cmc), both HPC and HPMC become fully extended and the lipophilic domains are encapsulated in surfactant micelles. HPMC enhances foam generation and strength as well, similar to MC. The lipophilic and surface-active HPMC also maintains the stability of o/w emulsions even after heat sterilization, providing potential applications in preparation of sterile creams and lotions.[39,76]

4.3.2.4.7 Ethylhydroxyethylcellulose (EHEC). Ethylhydroxyethylcellulose (EHEC) has the ability to gel at low temperatures slightly above 30 °C, especially in the presence of cationic and anionic surfactants.[77–79] As the temperature is raised, the hydrophobic domains on EHEC form strong inter-polysaccharide cross-links. This thermal gelation effect is influenced by the concentrations of EHEC and surfactants present in the solution. Currently, EHEC has limited applications in personal care but could potentially be useful in the delivery of therapeutics in "*cosmeceuticals*" in the future.[39,77]

4.3.2.4.8 Amphoteric Polysaccharides. Amphoteric polyglycans are capable of carrying both cationic and anionic charges on the same chain. These polysaccharides are still relatively unknown and underutilised in cosmetics, although amphoteric surfactants are quite commonly used in personal care such as soaps. The complexity of their ionic nature provides many challenges

Table 4.3 Structures of *N*-, *O*- and *N,O*-carboxymethylchitosan[39]

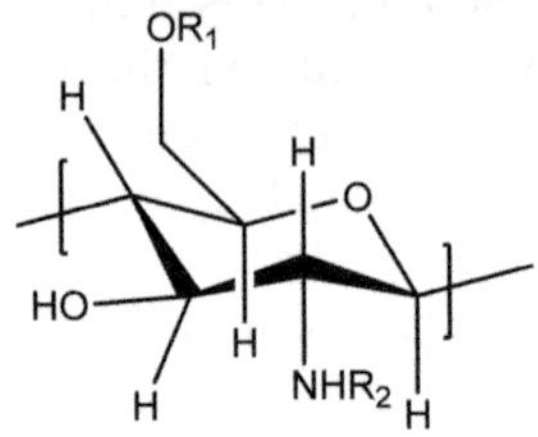

Name	R_1	R_2
N-Carboxymethylchitosan	H	CH_2COOH
O-Carboxymethylchitosan	CH_2COOH	H
N,O-Carboxymethylchitosan	CH_2COOH	CH_2COOH

in formulation such as solubility, pH stability and presence of salts or surfactants. Typically, trial and error is the only feasible method for figuring out the optimum formulation containing such polysaccharides.

4.3.2.4.9 Carboxymethylchitosan. Carboxymethylchitosan is the most well-known amphoteric polysaccharide with three available varieties: *N*-, *O*- and *N,O*-carboxymethylchitosan (Table 4.3). Among them, *N,O*-carboxymethylchitosan is the easiest to manufacture by a simple carboxymethylation reaction using chloroacetic acid. The main advantage of an amphoteric chitosan is improved water solubility. Chitosan is normally only soluble in water under acidic conditions, while carboxymethylchitosan is able to dissolve under alkaline conditions as well. However, carboxymethylchitosan is zwitterionic and water insoluble at a pH near 4.0. At extremely low pH values (*e.g.* 1.0), it becomes cationic and redissolves again.[39]

4.3.2.4.10 *N*-[(2′-Hydroxy-2′,3′-dicarboxy)ethyl]chitosan. Another amphoteric derivative can be synthesized from chitosan by reaction with anionic epoxides like *cis*-epoxysuccinic acid (Figure 4.15).[80] Similar to carboxymethylchitosan, it is water soluble at alkaline pH but phase separates at a pH near 4.0. The high anionic charge density allows it to, like chitosan, gel in the presence of various multivalent cations and cationic polymers. In comparison with natural chitosan, it exhibits good stability in the presence of anionic surfactants and polymers.[39]

4.3.2.4.11 Modified Potato Starch. An amphoteric ether can be derived from native potato starch with an amphoteric substituent.[81] Although the level of amphoteric substitution is low, a high concentration of anionic charge surrounds the tertiary nitrogen. The modified potato starch has good solubility over the entire pH range and helps to thicken formulations with extreme pH conditions, where native potato starch fails. In addition, at low pH levels, the amphoteric substituent protects the starch backbone from hydrolysis.[39]

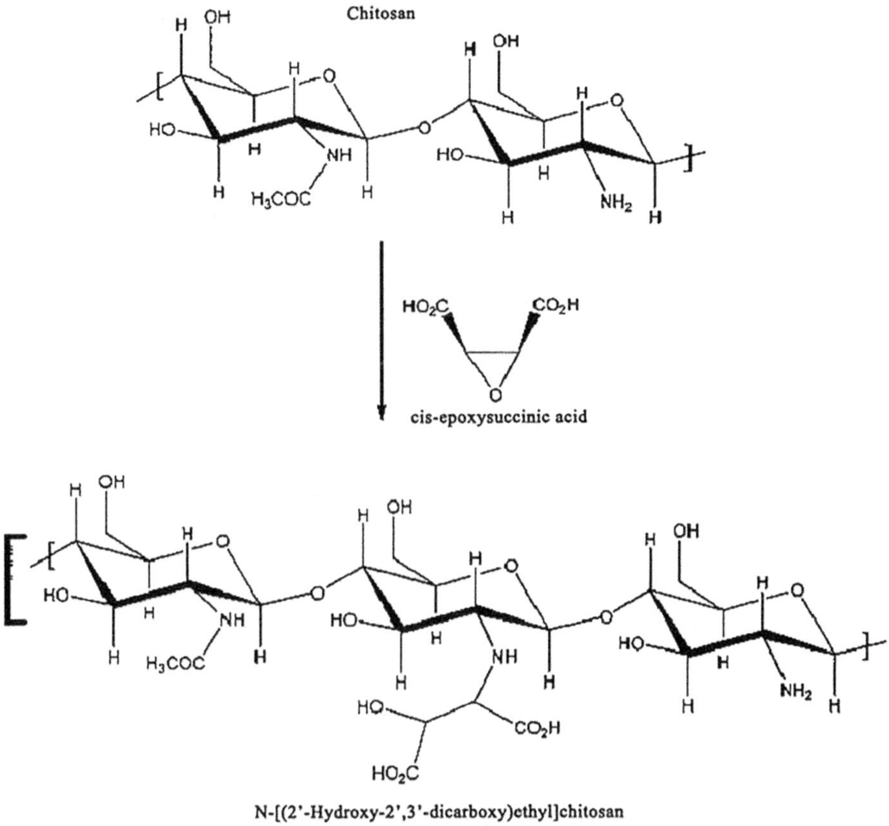

Figure 4.15 Synthesis of *N*-[(2′-hydroxy-2′,3′-dicarboxy)ethyl]chitosan from chitosan.

4.4 Future Outlooks

In closing, natural polysaccharides as rheological modifiers will continue to offer new exciting opportunities with the rising environmental and consumer concern over renewable sources of rheological modifiers and the origin of the ingredients used in personal care products.[1,4] Natural polysaccharides offer a wide variety of rheological modifiers to choose from as well as possible synthetic modifications to semi-natural derivatives with improved thickening effects and solution properties. As progress in genetic engineering advances, we look forward to many more interesting and novel polysaccharides that can be synthesized in laboratories.

References

1. S. Kumar, *Technovation*, 2005, **25**, 1263–1272.
2. D. Heard, *Putting On A Good Face - The Chemistry of Cosmetics*, 2015, http://www.sciencearchive.org.au/nova/083/083key.html, accessed 21 March 2015.

3. J. Castro, *Cosmetic Chemistry*, 2015, http://www.chemistryexplained.com/Co-Di/Cosmetic-Chemistry.html, accessed 17 March 2015.
4. G. M. Simmie, *Chemistry in the Modern World*, 2015, http://www.pharmachemicalireland.ie/Sectors/PCI/PCI.nsf/vPages/Education~Teachers~cosmetics-01-09-2009/$file/cosmetics%20module2.pdf, accessed 21 March 2015.
5. FDA, *Is It a Cosmetic, a Drug, or Both? (Or Is It Soap?)*, 2015, http://www.fda.gov/Cosmetics/GuidanceRegulation/LawsRegulations/ucm074201.htm, accessed 21 March 2015.
6. G. B. A. Judah, R. P. M. Aunger, W.-P. M. D. M. Schmidt, S. D. C. Michie, S. P. Granger and V. P. M. Curtis, *American Journal of Public Health*, 2009, **99**, S405–S411.
7. D. T. Jamison, J. G. Breman, A. R. Measham, G. Alleyne, M. Claeson, D. B. Evans, P. Jha, A. Mills and P. Musgrove, *Disease Control Priorities in Developing Countries*, The World Bank, 2nd edn, 2006.
8. J. Borghi, L. Guinness, J. Ouedraogo and V. Curtis, *Trop. Med. Int. Health*, 2002, **7**, 960–969.
9. V. Curtis and S. Cairncross, *Lancet Infect. Dis.*, 2003, **3**, 275–281.
10. T. Rabie and V. Curtis, *Trop. Med. Int. Health*, 2006, **11**, 258–267.
11. A. Biran, W.-P. Schmidt, R. Wright, T. Jones, M. Seshadri, P. Isaac, N. A. Nathan, P. Hall, J. McKenna, S. Granger, P. Bidinger and V. Curtis, *Trop. Med. Int. Health*, 2009, **14**, 1303–1314.
12. J. C. M. Heijne, P. Teunis, G. Morroy, C. Wijkmans, S. Oostveen, E. Duizer, M. Kretzschmar and J. Wallinga, *Emerging Infect. Dis.*, 2009, **15**, 24–30.
13. J. M. Lieberman, *Pediatr. Ann.*, 1994, **23**, 529–532.
14. M. L. Grayson, S. Melvani, J. Druce, I. G. Barr, S. A. Ballard, P. D. R. Johnson, T. Mastorakos and C. Birch, *Clin. Infect. Dis.*, 2009, **48**, 285–291.
15. M. A. Moyad and L. E. Robinson, *Urol. Nurs.*, 2008, **28**, 146.
16. L. J. Jarvie, R. Martin, P. D. R. Johnson, M. E. Jodoin, C. McMullan, R. H. C. Gregory, K. Bellis, K. Cunnington, F. L. Wilson, D. Quin and A.-M. Kelly, *Med. J. Aust.*, 2008, **188**, 633–640.
17. G. Judah, P. Donachie, E. Cobb, W. Schmidt, M. Holland and V. Curtis, *Epidemiol. Infect.*, 2010, **138**, 409–414.
18. A. Fallon, in *Body images: development, deviance, and change*, ed. T. F. Cash and T. Pruzinsky, Guilford Press, New York, 1990.
19. E. A. Saltzberg and J. C. Chrisler, in *Women: A Feminist Perspective*, ed. J. Freeman, Mayfield Publishing Company, Mountain View, 5th edn, 1995.
20. K. Dion, E. Berscheid and E. Walster, *J. Pers. Soc. Psychol.*, 1972, **24**, 285–290.
21. J. C. Larkin and H. A. Pines, *Work and Occupations*, 1979, **6**, 312–327.
22. E. D. Rothblum, C. T. Miller and B. Garbutt, *Int. J. Eating Disord.*, 1988, **7**, 277–283.
23. L. Karris, *J. Soc. Psychol.*, 1977, **101**, 159.
24. Statista, *Breakdown of the cosmetic market worldwide from 2011 to 2013, by product category*, http://www.statista.com/statistics/243967/breakdown-of-the-cosmetic-market-worldwide-by-product-category/, accessed 23 March 2015, 2015.

25. M. Yeomans, *Report estimates global personal care industry to reach €487 billion by 2017*, 2015, http://www.cosmeticsdesign-europe.com/Market-Trends/Report-estimates-global-personal-care-industry-to-reach-487-billion-by-2017, accessed 23 March 2015.
26. D. C. Duber-Smith, *Formulating, Packaging, and Marketing of Natural Cosmetic Products*, John Wiley & Sons, Inc., 2011, pp. 1–14, DOI: 10.1002/9781118056806.ch1.
27. M. Yeomans, *Global organic cosmetics market to reach $13.2 billion by 2018*, 2015, http://www.cosmeticsdesign-europe.com/Market-Trends/Global-organic-cosmetics-market-to-reach-13.2-billion-by-2018?utm_source=-copyright&utm_medium=OnSite&utm_campaign=copyright, accessed 23 March 2015.
28. S. Pitman, *Organic personal care market likely to post double-digit annual growth to 2020*, 2015, http://www.cosmeticsdesign-europe.com/Market-Trends/Organic-personal-care-market-likely-to-post-double-digit-annual-growth-to-2020, accessed 23 March 2015.
29. S. G. Mokrushin, *Nature*, 1962, **195**, 861.
30. P. Becher, Emulsions: theory and practice, Reinhold, New York U6 - ctx_ver-=Z39.88–2004&ctx_enc=info%3Aofi%2Fenc%3AUTF-8&rfr_id=info:sid/summon.serialssolutions.com&rft_val_fmt=info:ofi/fmt:kev:mtx:-book&rft.genre=book&rft.title=Emulsions&rft.au=Becher%2C+Paul&rft.series=American+Chemical+Society.+Monograph+series&rft.date=1957-01-01&rft.pub=Reinhold&rft.volume=no.+135.&rft.externalDo-cID=b12893341&paramdict=en-US U7 - Book, 1957.
31. *Soap, Water and Oil*, http://nsb.wikidot.com/c-9-5-5-4.
32. C. E. Ltd, *Span and Tween*, 2015, http://www.google.com.sg/url?sa=t&rct=j&q=&esrc=s&source=web&cd=1&cad=rja&uact=8&sqi=2&ved=0CBwQFjAA&url=http%3A%2F%2Fchemagent.ru%2Fcomponent%2Fflexi-content%2Fdownload%2F849%2F970%2F19&ei=NUgQVe_dI8S9ugTM-mIG4Cw&usg=AFQjCNGMuYgP3Ij8RPb8x06PMTtKKjd7OQ&sig2=6sZxJ-dJmY9hpB5t5-IOiRQ, accessed 24 March 2015.
33. R. M. Savage, *Food Hydrocolloids*, 2000, **14**, 209–215.
34. C. E. Jones, P. F. D. Reeve, United States Pat., 5,916,967, 1999.
35. E. Industries, *Rheological Additives: TEGO® ViscoPlus*, 2015, http://www.tego.de/sites/dc/Downloadcenter/Evonik/Product/Tego/en/Technical-Background/rheological-additives.pdf, accessed 23 March 2015.
36. Y. L. Robert, *Cosmetic Nanotechnology*, American Chemical Society, 2007, ch. 1, vol. 961, pp. 3–56.
37. B. SE, *Practical Guide to Rheology Modifiers*, 2015, https://www.dispersions-pigments.basf.com/portal/load/fid793184/BASF%20Rheology%20Modifiers%20Practical%20Guide.pdf, accessed 11 March 2015.
38. T. O. Olaleye, *SPC/SCS essay prize - gums, thickeners & resins*, 2015, http://www.cosmeticsbusiness.com/technical/article_page/SPCSCS_essay_prize__gums_thickeners__resins/56082, accessed 17 March 2015.
39. V. G. James, *Principles of Polymer Science and Technology in Cosmetics and Personal Care*, CRC Press, 1999, DOI: 10.1201/9780203907948.ch8.

40. G. Sworn, *Handbook of Industrial Water Soluble Polymers*, Blackwell Publishing Ltd., 2007, pp. 10–31, DOI: 10.1002/9780470988701.ch2.
41. E. R. Morris, A. N. Cutler, S. B. Ross-Murphy, D. A. Rees and J. Price, *Carbohydr. Polym.*, 1981, **1**, 5–21.
42. M. Inc, *Polysaccharides (Sugars, Gums) Used in Cosmetics*, 2015, http://www.makingcosmetics.com/articles/20-polysaccharides-sugars-gums-in-cosmetics.pdf, accessed 17 March 2015.
43. R. L. Davidson, *Handbook of water-soluble gums and resins*, McGraw-Hill, New York, 1980.
44. K. Clare, in *Industrial Gums*, ed. R. L. Whistler and J. N. Bemiller, Academic Press, London, Third edn, 1993, pp. 105–143, DOI: 10.1016/B978-0-08-092654-4.50010-3.
45. M. Yalpani, *Polysaccharides: syntheses, modifications, and structure/property relations*, Elsevier, Amsterdam, New York, 1988.
46. R. E. Kirk, D. F. Othmer, J. I. Kroschwitz and M. Howe-Grant, *Encyclopedia of chemical technology*, Wiley, New York, 1991.
47. K. I. Draget, G. Skjåk-Bræk and O. Smidsrød, *Int. J. Biol. Macromol.*, 1997, **21**, 47–55.
48. J. W. Lee, R. D. Ashby and D. F. Day, *Carbohydr. Polym.*, 1996, **29**, 337–345.
49. B. T. Stokke, O. Smidsroed, P. Bruheim and G. Skjaak-Braek, *Macromolecules*, 1991, **24**, 4637–4645.
50. M. Djabourov, *Polym. Int.*, 1991, **25**, 135–143.
51. A. H. Clark, K. Kamide, S. B. Ross-Murphy and M. Saito, *Advances in Polymer Science - 83 Biopolymers*, Springer-Verlag, Heidelberg, 1987.
52. D. Thom, G. T. Grant, E. R. Morris and D. A. Rees, *Carbohydr. Res.*, 1982, **100**, 29–42.
53. C. Rolin, *Industrial Gums*, ed. R. L. Whistler and J. N. Bemiller, Academic Press, London, Third edn, 1993, pp. 257–293, DOI: 10.1016/B978-0-08-092654-4.50014-0.
54. R. Chandrasekaran and A. Radha, *Carbohydr. Polym.*, 1997, **32**, 201–208.
55. G. H. Therkelsen, in *Industrial Gums*, eds. R. L. Whistler and J. N. Bemiller, Academic Press, London, Third edn, 1993, pp. 145–180, DOI: 10.1016/B978-0-08-092654-4.50011-5.
56. R. J. Tye, *Drug & Cosmetic Industry*, 1991, ch. 28, vol. 149, p. 28+.
57. D. B. Braun and M. R. Rosen, *The rheology modifiers handbook: practical use & application*, Noyes Publications/William Andrew Pub, Norwich, NY, 1999.
58. C. Rochas, M. Rinaudo and S. Landry, *Carbohydr. Polym.*, 1990, **12**, 255–266.
59. R. K. Richardson and F. M. Goycoolea, *Carbohydr. Polym.*, 1994, **24**, 223–225.
60. A. M. Hermansson, E. Eriksson and E. Jordansson, *Carbohydr. Polym.*, 1991, **16**, 297–320.
61. E. R. Morris, D. A. Rees and G. Robinson, *J. Mol. Biol.*, 1980, **138**, 349–362.
62. L. Piculell, J. Borgström, I. S. Chronakis, P. O. Quist and C. Viebke, *Int. J. Biol. Macromol.*, 1997, **21**, 141–153.
63. T. Turquois, C. Rochas and F. R. Taravel, *Carbohydr. Polym.*, 1992, **17**, 263–268.

64. R. L. Feddersen and S. N. Thorp, in *Industrial Gums*, ed. R. L. Whistler and J. N. Bemiller, Academic Press, London, Third edn, 1993, pp. 537–578, DOI: 10.1016/B978-0-08-092654-4.50024-3.
65. Hercules, *Aqualon® Sodium Carboxymethylcellulose: Physical and Chemical Properties*, 2015, http://www.brenntagspecialties.com/en/downloads/Products/Multi_Market_Principals/Aqualon/Aqualon_CMC_Booklet.pdf, accessed 24 March 2015.
66. G. W. Gooday, M. Aruchami, N. Gowri and G. Sundara-Rajulu, in *Chitin in Nature and Technology*, ed. R. Muzzarelli, C. Jeuniaux and G. Gooday, Springer US, 1986, ch. 34, pp. 263–268, DOI: 10.1007/978-1-4613-2167-5_34.
67. M. F. A. Goosen, *Applications of Chitin and Chitosan*, Technomic Pub, Lancaster, PA, 1997.
68. P. S. Leung, E. D. Goddard, C. Han and C. J. Glinka, *Colloids Surf.*, 1985, **13**, 47–62.
69. E. D. Goddard and R. B. Hannan, in *Micellization, Solubilization, and Microemulsions*, ed. K. L. Mittal, Springer US, 1977, ch. 20, pp. 835–845, DOI: 10.1007/978-1-4613-4157-4_20.
70. M. Goldraich, J. R. Schwartz, J. L. Burns and Y. Talmon, *Colloids Surf., A*, 1997, **125**, 231–244.
71. M. Malmsten, I. Ljusegren and I. Carlstedt, *Colloids Surf., B*, 1994, **2**, 463–470.
72. N. Beşün, S. Peker, U. Köktürk and H. Yilmaz, *Colloid Polym. Sci.*, 1997, **275**, 378–389.
73. F. F. Mikus, R. M. Hixon and R. E. Rundle, *J. Am. Chem. Soc.*, 1946, **68**, 1115–1123.
74. A. J. Desmarais and R. F. Wint, in *Industrial Gums*, ed. R. L. Whistler and J. N. Bemiller, Academic Press, London, Third edn, 1993, pp. 505–535, DOI: 10.1016/B978-0-08-092654-4.50023-1.
75. J. A. Grover, in *Industrial Gums*, ed. R. L. Whistler and J. N. Bemiller, Academic Press, London, Third edn, 1993, pp. 475–504, DOI: 10.1016/B978-0-08-092654-4.50022-X.
76. K. Hayakawa, M. Kawaguchi and T. Kato, *Langmuir*, 1997, **13**, 6069–6073.
77. A. Carlsson, G. Karlström and B. Lindman, *Colloids Surf.*, 1990, **47**, 147–165.
78. B. Nystroem and B. Lindman, *Macromolecules*, 1995, **28**, 967–974.
79. B. Nyström, A.-L. Kjøniksen and B. Lindman, *Langmuir*, 1996, **12**, 3233–3240.
80. J. V. Gruber, V. Rutar, J. Bandekar and P. N. Konish, *Macromolecules*, 1995, **28**, 8865–8867.
81. S. Cardinali, J. Greenberg and J. Pasapane, *Drug & Cosmetic Industry*, 1997, ch. 42, vol. 160, p. 42+.

CHAPTER 5

Antibacterial Polymers

HWA YAW JONATHAN HENG[b] AND XIAN JUN LOH*[a,b]

[a]Institute of Materials Research and Engineering, A*STAR (Agency for Science, Technology and Research), 2 Fusionopolis Way, Innovis, #08-03, Singapore 138634, Singapore; [b]Department of Materials Science and Engineering, National University of Singapore, Singapore 117574, Singapore
*E-mail: lohxj@imre.a-star.edu.sg

5.1 Introduction

Bacterial infections are an important concern in the medical field. They can be found almost everywhere on earth; from the air, to the ground, to plants and animals as well as in water. Even though most types of bacteria are not harmful to humans, some types of bacteria can cause infections. Bacteria reproduce quickly in our bodies and produce toxins that damage human tissue. Bacterial infections can lead to diseases such as pneumonia. The prevention of bacterial infections has been one of the most important considerations in the field of medical services and healthcare, where in some cases, infections can easily lead to death.

Bacteria are microorganisms that are a few micrometers in size. A bacterial cell mainly consists of the cell membrane and cell wall. The cell wall is important for the cell as it provides support and protection. In general, bacteria can be classified into two main types due to the differences in their cell walls. Gram-positive bacteria have a thicker cell wall, made of peptidoglycans, while gram-negative bacteria have a thinner cell wall. However,

RSC Polymer Chemistry Series No. 20
Polymers for Personal Care Products and Cosmetics
Edited by Xian Jun Loh

Published by the Royal Society of Chemistry, www.rsc.org

gram-negative bacteria have an additional outer membrane layer, which makes them more resistant and harder to kill as compared to gram-positive bacteria (Figure 5.1).[1]

Our immune system is able to protect the body during an infection. People with weaker immune systems such as infants or the elderly are more susceptible to infection. Antibiotics can also be used as a treatment to kill bacteria inside the body. However, some bacterial strains are able to mutate and develop a resistance to current antibiotics, hence there is a need to find another method of killing bacteria more effectively. There has been great development in the area of antibacterial polymers as antimicrobial agents. Antibacterial polymers are polymers that possess antibacterial properties that are able to inhibit bacterial growth, or even kill bacteria. Antibacterial polymers can be classified into 3 different groups: (1) polymers with inherent antibacterial activity, (2) chemically modified polymers, (3) addition of antibacterial agents.

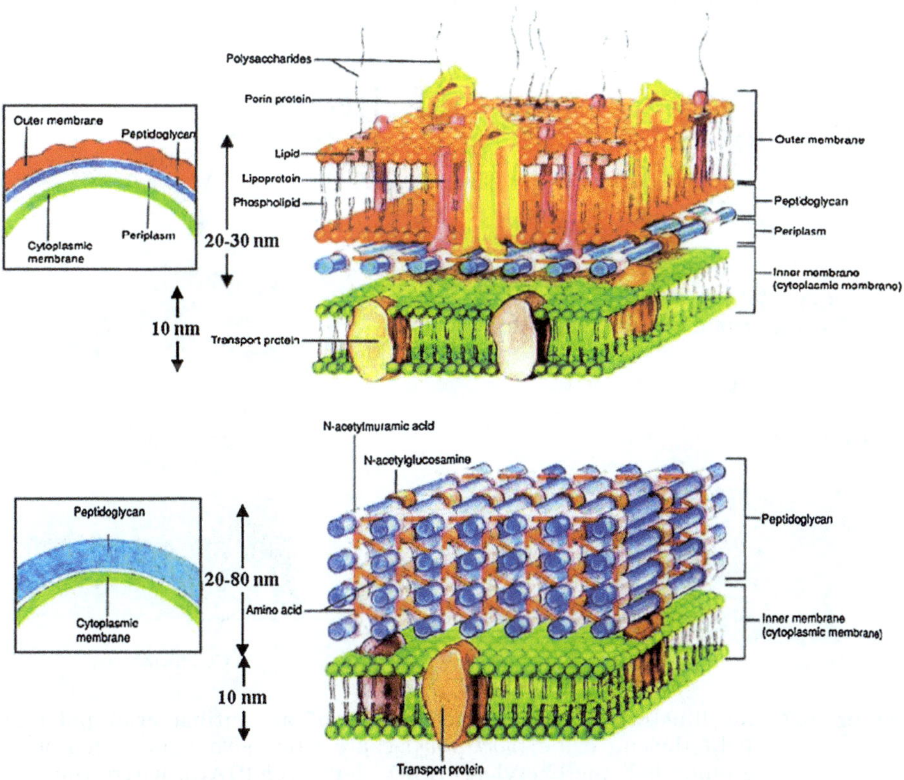

Figure 5.1 Schematic views of gram-negative (top) and gram-positive (bottom) bacterial cell walls, showing the complex assembly of high molecular weight compounds. Reproduced from ref. 1 with permission from the Royal Society of Chemistry.

5.1.1 Antibacterial Polymers and Their Mechanism

Antibacterial polymers are generally designed as cationic macromolecules that have both hydrophilic and hydrophobic interactions. These polymers provide an active surface as well as good adsorption and absorption properties, which allow favorable binding and interaction to occur with bacterial cells.[2] In general, antibacterial polymers kill bacteria when the bacterial cells come into contact with the surface of the polymer or its molecules. As the polymers diffuse through the cell walls, adsorption onto the plasma membrane of the cell occurs, which disrupts the membrane.[3,4] This results in the leakage of the membrane, leading to cell death (Figure 5.2).

Gram-negative bacterial cells, such as *Pseudomonas aeruginosa* (*P. aeruginosa*) and *Escherichia coli* (*E. coli*), have an additional outer membrane, as compared to gram-positive bacterial cells, such as *Staphylococcus aureus* (*S. aureus*). This makes gram-negative bacteria more resistance to biocides such as antibiotics and antibacterial polymers. Most of the antibacterial polymers in this review are tested against *E. coli* and *S. aureus* as they are very common bacterial infections in humans.

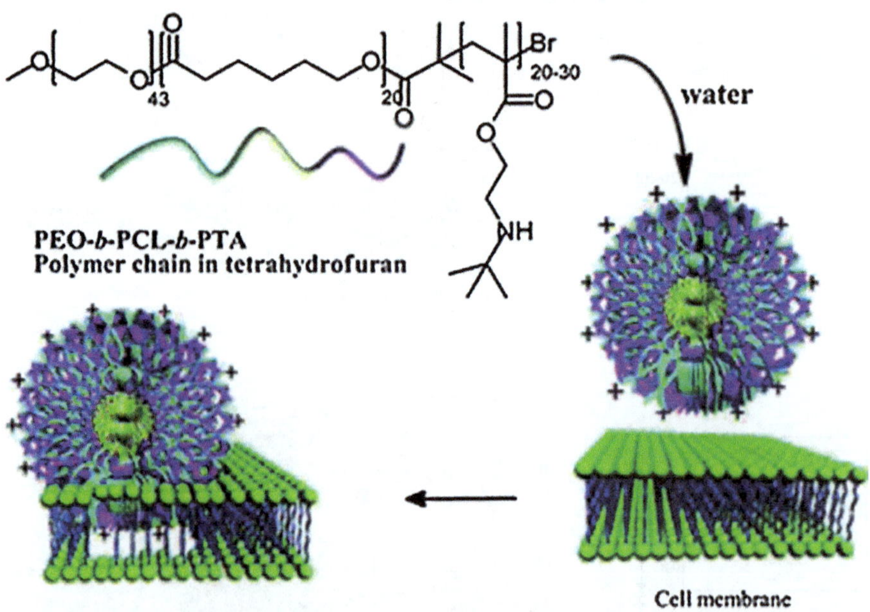

Figure 5.2 An illustration of the mechanism of an antibacterial polymer. Poly(ethylene oxide)-*block*-poly(ε-caprolactone)-*block*-poly[(2-*tert*-butylaminoethyl) methacrylate] (PEO_{43}-*b*-PCL_{20}-*b*-PTA_{20}), where the subscripts denote the degree of polymerization. The biodegradable PCL is engineered to drive the copolymers to self-assemble into micelles. PTA is designed to interact with the microbial wall/membrane. Reproduced from ref. 4 with permission from the Royal Society of Chemistry.

The resistance of the antibacterial polymers can be determined by the quantities of minimum inhibitory concentration (MIC) and minimum bactericidal concentration (MBC). The MIC value determines the lowest concentration of antibacterial agent that inhibits bacterial growth from its initial inoculum. The MBC value is determined by the lowest concentration of antibacterial agent that reduces the viability of the initial bacterial inoculum. Antibacterial agents are usually regarded as bactericidal, *i.e.* able to kill bacteria, if the MBC is no more than four times the MIC.

5.1.2 History of Antibacterial Polymers

One of the earliest discoveries of antibacterial polymers was the finding by Cornell and Donaruma, who found that the bacterial activity decreased[5] when comparing the monomer 2-methacryloxytropone and its polymer with several bacteria strains (Table 5.1).

Today, many functional groups or agents are well known to possess antibacterial capabilities. Many antibacterial polymers are cationic, such as quaternary nitrogen groups, sulfonium salts and phosphonium groups. Nitrogen-containing compounds such as polyguanidines and *N*-halamines are well known for their antibacterial activity. Phenol groups are also known for their antibacterial properties.

It has become increasingly more important to develop new and more effective antibacterial polymers today. Bacteria are able to develop resistance to antibiotics. The methicillin-resistant *Staphylococcus aureus* (MRSA) strain is just one example. Antibacterial polymers have become a potential alternative biocide, as bacterial cells are not able to develop immunity towards them. Antibacterial polymers are widely used in many areas of application today. Bacteria are found almost everywhere, and there are many applications in which minimal or no bacterial growth is ideal. Many companies, such as Microban, Parx Plastics and RTP Company, provide antibacterial plastics that can be used for a wide variety of surfaces such as aircraft interiors, baby changing stations, food processing and many others. Current research is ongoing to develop antibacterial polymers for textiles, fabrics, water treatment, food packaging, medical devices and healthcare. Research in these areas has shown promising results.

Table 5.1 Comparison of antibacterial activity of 2-methacryloxytropone and its polymer.

Bacterial species	Monomer activity[a]	Polymer activity[a]
Staphyloccocus aureus no. 6538	15	12
Salmonella typhosa no. 6539	22	8
Salmonella choleraesuis no. 10708	17	8
Escherichia coli no. 11229	16	10
Streptococcus pyogenes no. 624	17	14

[a]Width of zone of growth inhibition in mm.

5.2 Polymers with Inherent Antibacterial Activity

5.2.1 Chitosan

Chitosan is a natural polysaccharide derived from chitin, which is present in the exoskeletal shells of crustaceans such as shrimps and crabs. Chitosan is formed from chitin by deacetylation of chitin. Allan and Hadwiger first discovered the antibacterial activity of chitosan[6] in 1979, where they found that the MIC value for chitosan was lower in general when compared to chitin. This meant that chitosan had stronger antibacterial activity than chitin. The intrinsic antibacterial property of chitosan was shown to be effective against both gram-positive as well as gram-negative bacteria. However, there are conflicting research findings as to which type of bacteria chitosan is more effective against. In 2008, one research group reported that chitosan was more effective against gram-positive bacteria due to the thick outer membrane of gram-negative bacteria,[7] while in 2004, another research group had found that more chitosan absorbed onto gram-negative bacteria,[8] and that it was also more effective in killing gram-negative bacteria than gram-positive bacteria. The antibacterial activity of chitosan increases when dissolved. As chitosan becomes dissociated when dissolved in solution, it adopts an extended conformation which increases the effectiveness of its antibacterial activity.[9] This has led to further research to synthesize new water-soluble chitosan compounds. Some examples include chitosan-glucosamine derivatives[10] and quaternary ammonium chitosan derivatives,[11] which both show improved solubility and more effective antibacterial properties as compared to chitosan itself. Current research on chitosan includes many areas of application such as biomedical, textiles and food packaging. Other than its antibacterial properties, this interest in research on chitosan in these areas is due to the fact that chitosan is also biocompatible, non-toxic, biodegradable, environmentally friendly and able to be spun into fibers. Chitosan is also a well known food additive that helps preserve food, which serves to prevent bacterial growth and improve shelf life. It can also be used in food packaging material, including biodegradable films.[12] As chitosan can be processed in many different forms, such as hydrogels, membranes or fibers, it can also be used for wound dressings.[13] Chitosan can also be used as a finishing on textiles to prevent bacterial growth as well as prevent bad odor on clothing. Carboxymethyl chitosan was coated onto cotton fabric, displaying good antibacterial properties.[14] Chitosan-based compounds can also be used to fabricate scaffolds and matrices for cells to grow on.[15]

5.2.2 Quaternary Ammonium Polymers

Polymers containing quaternary ammonium groups are one of the most widely researched antibacterial polymers. These polymers have greater interaction with gram-positive bacterial cells as compared to gram-negative cells, as the polymer is able to penetrate more deeply into the gram-positive cell structure. In addition, gram-negative bacterial cells have an additional outer membrane layer that is able to protect the cell better against the quaternary ammonium groups. One example of these polymers would be poly(2-(dimethylaminoethyl)methacrylate) (pDMAEMA). pDMAEMA displayed good antibacterial activity against a wide spectrum of both gram-positive and gram-negative bacterial strains (Table 5.2).

Also, the synthesis of quaternary ammonium salt monomers from dimethylaminoethyl methacrylate (DMAEMA) by quaternization with benzyl chloride (BC), butyl bromide (BB), dodecyl bromide (DB) or hexadecyl bromide (HB)[16] were all found to have antibacterial effects against *S. aureus* and *E. coli,* with poly(DMAEMA-BC) showing the strongest antibacterial activity. By comparing the antibacterial properties of the monomer and its polymer, they found that poly(DMAEMA-BC) and poly(DMAEMA-BB) had greater antibacterial effects as compared to their monomers, while poly(DMAEMA-DB) and poly(DMAEMA-HB) showed the opposite trend (Table 5.3).

Other cationic polymers that have quaternary ammonium groups include polymers with pyridine groups. These polymers have been developed since as early as 1988.[17] Four polymers: 4-aminododecylpyridinium chloride, 4-acetylaminododecylpyridinium chloride, 4-benzoylaminododecylpyridinium bromide and 4-(1-naphthoyl)aminododecylpyridinium bromide were

Table 5.2 MIC concentrations (mg ml^{-1}) for pDMAEMA against a range of bacterial strains and yeast.

Organism	Gram	MIC
E. coli ATCC 10536	–	0.1
E. coli equine isolate	–	0.1
Salmonella ser. Enteritidis ATCC 13076	–	0.1
P. aeruginosa QC strain	–	1
Micrococcus luteus ATCC 9341	+	>18
Lactobacillus salivarius UCC 118	+	>18
Listeria monocytogenes NCTC 11994	+	10
Listeria spp. wild type #28	+	1
Bifidobacterium breve DSMZ 20213	+	10
Bifidobacterium bifidum DSMA 20456	+	10
Staphylococcus epidermidis 1457	+	0.1
Candida albicans C	Yeast	>10

Table 5.3 MBC values of quarternary ammonium salts monomers and their polymers.

	DMAEMA-BC		DMAEMA-BB		DMAEMA-DB		DMAEMA-HB	
Sample	Monomer (mg mL^{-1})	Polymer (mg mL^{-1})	Monomer (mg mL^{-1})	Polymer (mg mL^{-1})	Monomer (mg mL^{-1})	Polymer (mg mL^{-1})	Monomer (mg mL^{-1})	Polymer (mg mL^{-1})
E. coli	>50	1.56	>50	1.56	24	>20	12	>20
S. aureus	>50	1.56	>50	1.56	12	>20	12	>20

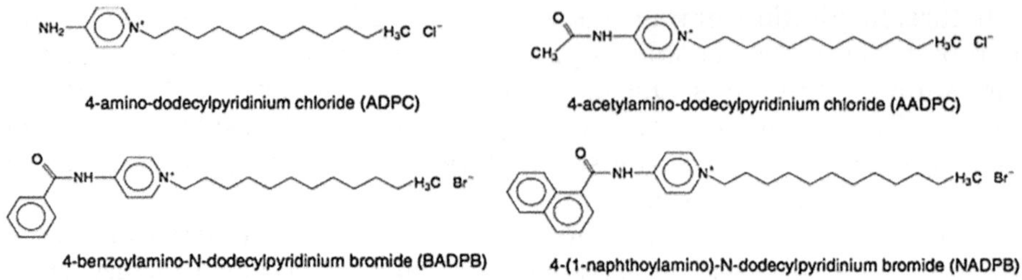

Figure 5.3 Structures of the 4 polymers with pyridine groups.

tested against *E. coli* and *S. aureus*. Despite all 4 polymers having the same alkyl chain length, they differed by the different hydrophobic groups at the other end of the pyridine groups (Figure 5.3).

All 4 polymers displayed good antibacterial activity, however their activities differed due to the different hydrophobic groups. Larger hydrophobic groups (*i.e.* BADPB and NADPB) resulted in stronger antibacterial activity.[18]

5.2.3 Hyperbranched Polymers

Hyperbranched or dendritic polymers also possess antibacterial activity. They are able to solubilize in water despite their highly branched configuration and at the same time able to adsorb bacterial membrane. The overall cationic charge of the polymer also contributes to the antibacterial effect due to the amino groups on the macromolecular chain, which cause cell death by disrupting the bacterial cell membrane. One example of a highly branched polymer is branched polyethylenimine (PEI). Quaternized PEI is known to possess antibacterial activity due to its cationic charge[19] and has been tested to be effective against *E. coli* and *S. aureus*. Further research has also been conducted on the synthesis of quaternary ammonium PEI-based nanoparticles[20] as potential additives in composites that require antibacterial properties in applications such as dental composites and antibacterial surfaces.

Antibacterial dendrimers available include quaternary ammonium polypropyleneimine (PPI) and polyamidoamine (PAMAM), which are commonly used in drug and DNA delivery. Quaternary ammonium PPI dendrimers (Figure 5.4) were synthesized, and their antibacterial activity was found to be related to the size of the polymer, as well as the length of the hydrophobic chains.[21]

Researchers used a bioluminescence method to evaluate the antibacterial activity of the polymers against *E. coli* and *S. aureus,* in which a lower relative bioluminescence value represented a stronger antibacterial activity. In general, quaternary ammonium polymers tend to be more effective against gram-positive bacteria such as *S. aureus*, as compared to gram-negative bacteria such as *E. coli*, due to the outer membrane structure of the gram-negative bacterial cells, which acts as an additional barrier of protection for

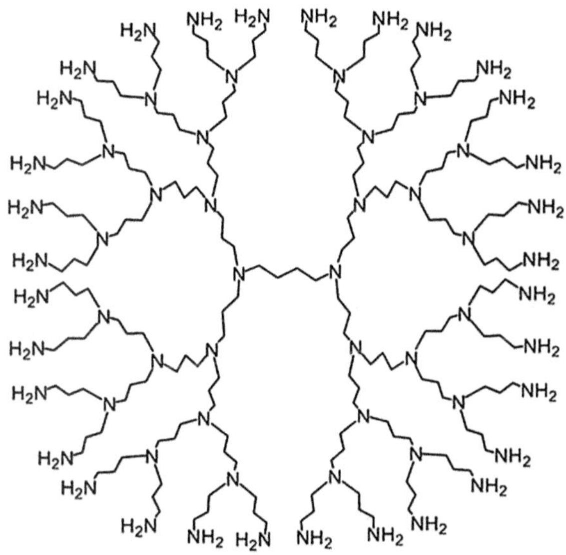

Figure 5.4 An example of a commercially available PPI dendrimer.

the cell. This makes it less susceptible to cell membrane disruption by other molecules such as cationic antibacterial polymers.

In a research study,[22] PAMAM (Figure 5.5) was found to be effective against *S. aureus* and *P. aeruginosa*.

Also, adding a coating of PEG was found to reduce PAMAM's cytoxicity when tested with human corneal epithelial cells (HCECs) while maintaining its antibacterial effect against the gram-negative *P. aeruginosa* species, but resulting in a large decrease in toxicity to the gram-positive *S. aureus* species. However, the drawbacks of these highly branched and dendritic polymers include the difficulty in controlling the homogeneity of the size of polymer molecules (polydispersity), which in turn makes it difficult to predict the effectiveness in disrupting the bacterial membranes. In addition, the amount of branching needs to be carefully considered, as the bulkiness of a polymer tends to restrict its ability to disrupt the bacterial membrane.

5.2.4 Polymers Containing Guanidine Groups

Guanidine is a functional group that contains the formula $HNC(NH_2)_2$. Guanidine and biguanidine derivatives have been known to possess antibacterial activity. Some guanidine and biguanidine salts were tested against a wide range of gram-positive bacteria and gram-negative bacteria, as well as some yeasts and fungi. Four polymers were synthesized: polyhexamethylene guanidine hydrochloride (PHGC), polyhexamethylene biguanidine hydrochloride (PHBGC), polyhexamethylene guanidine stearate (PHGS) and polyhexamethylene biguanidine stearate (PHBGS). These polymers have proven useful

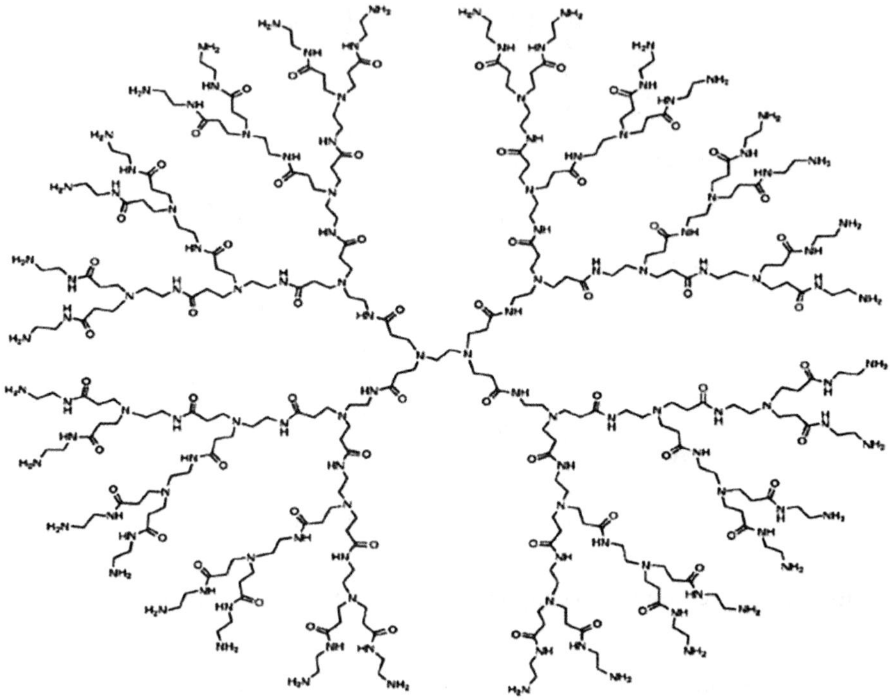

Figure 5.5 General structure of the PAMAM dendrimer.

as they exhibit other useful properties such as water solubility and thermal stability.

Despite their MIC values varying across the different strains of bacteria, they still generally show strong antibacterial activity (Table 5.4).

5.2.5 Polymers That Mimic Antimicrobial Peptides

Antimicrobial peptides such as magainin and cathelicidin (LL37) possess strong antibacterial activity and are also resistant to other microorganisms such as viruses and fungi. These naturally occurring peptides are produced by all species such as mammals, insects, fungi and even bacteria. They serve as host defense, as part of the organism's immune system. These synthetic polymers have been produced to mimic the properties of the natural antimicrobial peptides. There have been attempts to develop synthetic peptides from as early as 1988,[23] where the antimicrobial peptide magainin was isolated from frog skin. These natural peptides displayed strong antibacterial activity against a wide range of gram-positive and gram-negative bacteria. Synthetic peptides with the exact same structure as the natural peptides were then synthesized and shown to have identical antibacterial activity as their natural counterparts. The backbone chains of the antimicrobial peptides are able to disrupt bacterial cell membranes, which results in rapid cell death.

Current polycationic antibacterial polymers are actually very similar to magainin (Figure 5.6). The only main difference would be the stiff backbone structure of the peptide. This could provide an explanation why antimicrobial peptides generally have stronger antibacterial properties than most antibacterial polymers.

Since then, many other polymers have been developed to mimic the mechanism of these antimicrobial peptides. Poly(phenylene ethynylene)-based polymers are one common example of this.[25,26] These polymers mimic the structure and biological properties of antimicrobial peptides, showing comparable antibacterial properties against bacteria such as *E. coli* and *S. aureus.*

Table 5.4 MIC values (μg ml^{-1}) of the polymeric guanidine and biguanidine salts.

Samples	PHGC	PHGS	PHBGC	PHBGS
Bacillus subtilis	1.55	0.78	6.25	50
Sarcina	1.55	0.78	12.5	50
Staphylococcus aureus	<0.39	<0.39	12.5	0.78
Streptococcus pneumomiae	1.55	0.78	25	6.25
Escherchia	12.5	25	100	200
Pseudomonas aeruginosa	3.12	3.12	100	50
Rhizopus niger	0.78	3.12	12.5	12.5
Aspergillus niger	6.25	6.25	12.5	12.5
Saccharomyces cerevisiae	0.78	1.55	6.25	1.55
Candida albicans	0.39	0.78	<0.39	1.55

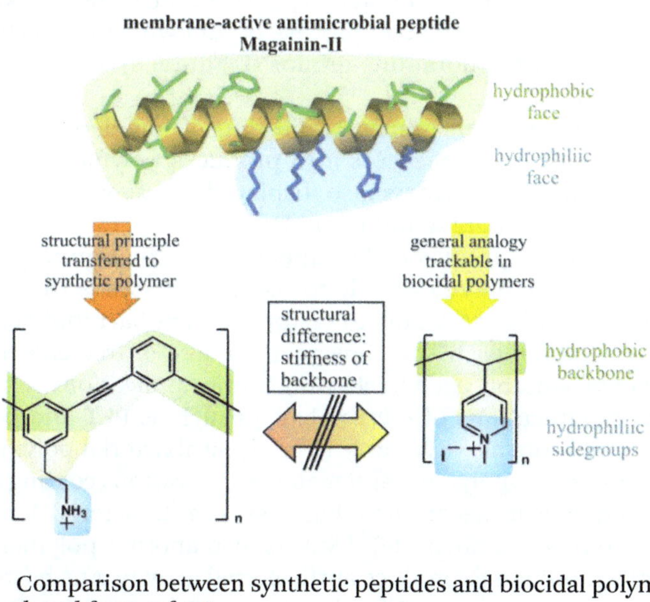

Figure 5.6 Comparison between synthetic peptides and biocidal polymers. Reproduced from ref. 24.

Also, several aryl amide compounds were synthesized[27] with varying side groups to analyze their molecular weights, structure and hydrophobicity, in relation to their antibacterial activity and toxicity to red blood cells.

These polymers displayed significant antibacterial activity against *E. coli* and *S. aureus*. However, compared to natural antimicrobial peptides, these synthesized polymers are shown to be toxic to human red blood cells at similar concentrations.

5.3 Chemically Modified Polymers

One major advantage of using antibacterial polymers is the ability to customize and fine-tune the antibacterial effects as well as other properties of the polymer. In this way, polymers that do not display any antibacterial activity are able to exhibit this desired property by grafting or coupling parts of antibacterial polymers to the inactive polymer. However, we do need to ensure that other properties are not greatly affected after this chemical modification.

5.3.1 *N*-Halamine-Based Groups

One example of this is the antibacterial treatment of nylon. Cyclic *N*-chloramine groups are known for their antibacterial properties and were covalently bonded to Nylon 6-6 (Figure 5.7). The antibacterial properties of the nylon fibers were then tested against *S. aureus* and *E. coli* and were found to be bactericidal.[28]

We are able to synthesize antibacterial polyester in the same way. *N*-Halamine groups were chemically bonded to polyester (PET). By exposing the new polymer to chlorine, an oxidative process occurs which converts the newly attached group into *N*-chloramine groups (Figure 5.8), which possess antibacterial properties.[29]

Recent research on melt graft modification of low density polyethylene (LDPE) with *N*-halamine precusors also produces polymers with a similar effect (Figure 5.9). The polymers also displayed strong antibacterial activity when tested against *S. aureus* and *E. coli*.[30]

As the *N*-halamine groups need to undergo chlorination to produce antibacterial activity, the only downside to these processes is the need to constantly introduce chlorine atoms to restore the antibacterial activity of the polymers. However, bleach is a convenient source of free chlorine and the polymers can be suitably used in clothing fabric applications.

Chemical modifications of inert polymers such as PET and polyethylene are not very common, as we are able to apply antibacterial polymers directly onto surfaces of other polymers and materials *via* various coating methods to produce different coatings or films that possess antibacterial activity.

Poly(styrene maleic anhydride) (SMA) is also another polymer which can be chemically modified to produce antibacterial activity. Antibacterial agents can be chemically bonded to SMA *via* a ring-opening reaction (Figure 5.10).

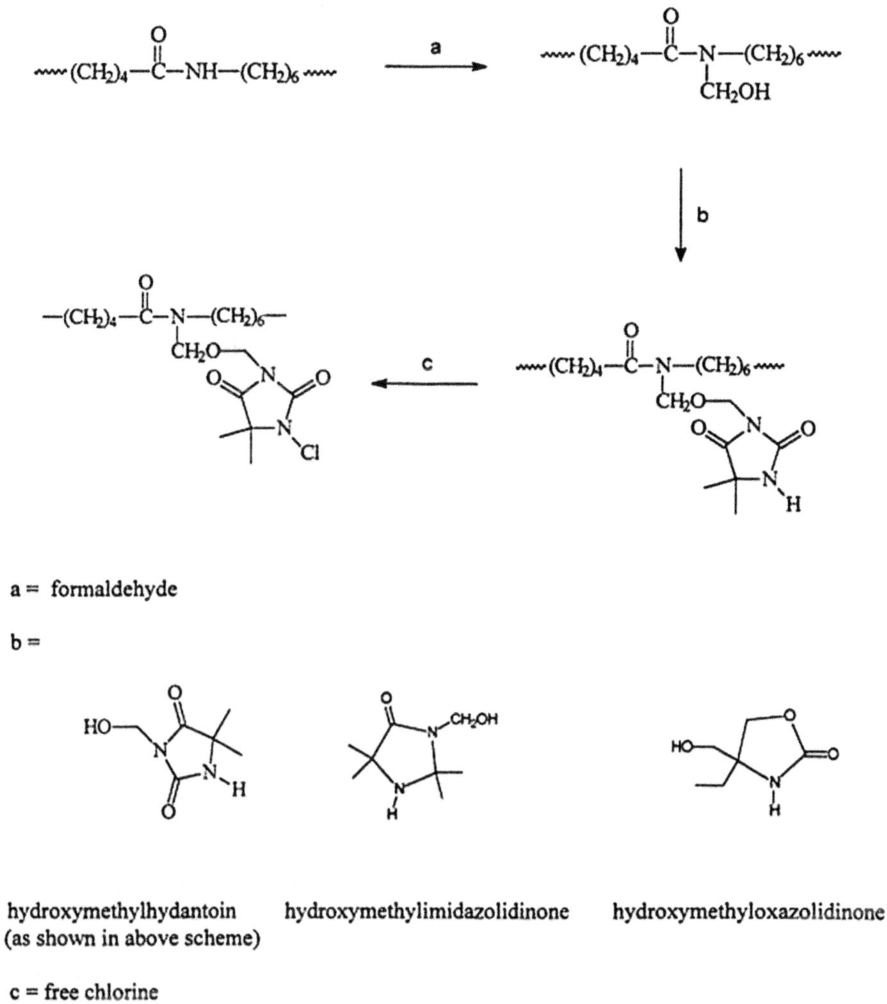

Figure 5.7 Reaction scheme used in producing the antimicrobial nylon samples.

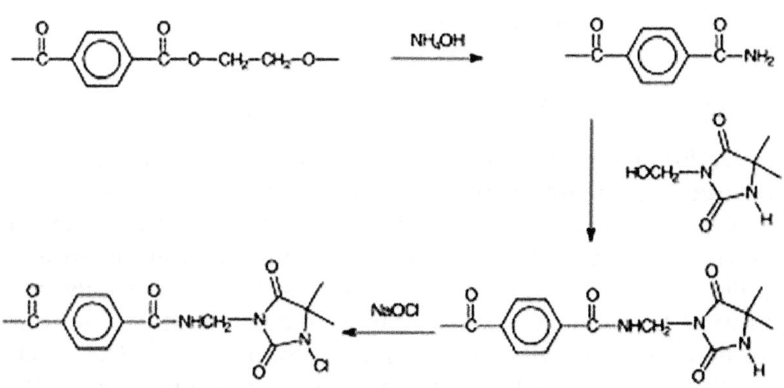

Figure 5.8 The reaction sequence used in producing the biocidal PET fabric.

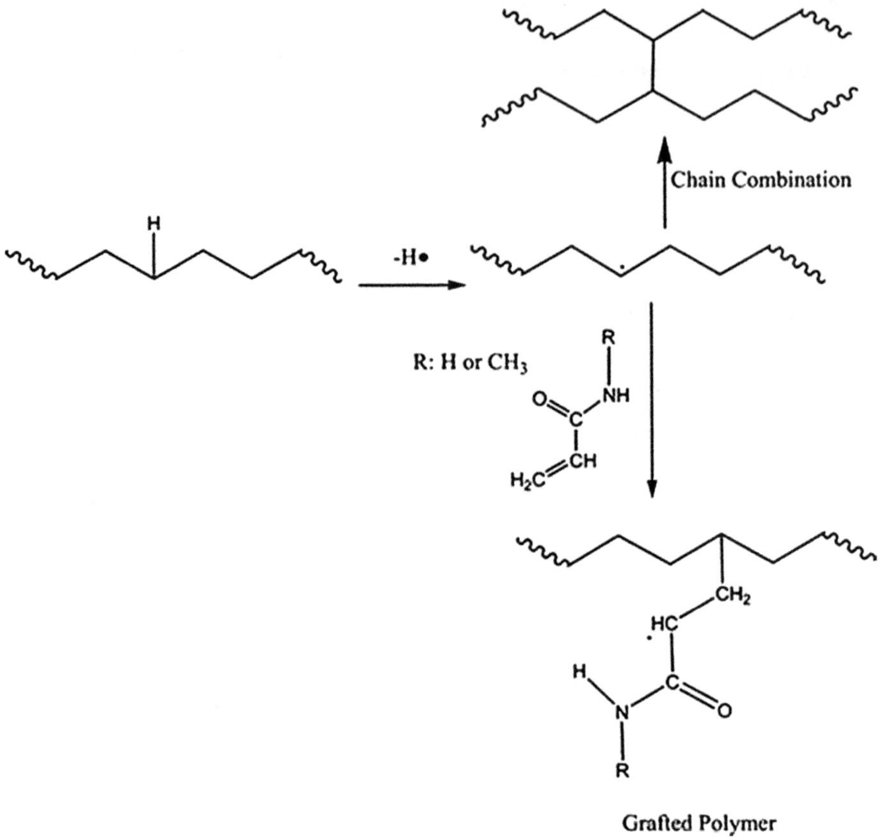

Figure 5.9 Overall grafting reaction of PE with acyclic halamine precursors in a reactive extrusion process.

4-Aminophenol (AP) is known to possess antibacterial activity due to its phenol group. The resultant polymer, SMA–AP was found to possess strong bactericidal activity against *E. coli* and *S. aureus.*[31] *Other polymers of modified SMA include SMA-bound* acriflavine,[32] SMA-bound ampicillin,[33] as well as SMA reacted with 4-hydroxybenzoic acid.[34]

5.4 Addition of Antibacterial Agents

Another method of synthesizing polymers with antibacterial activity would be to produce nanocomposites by using metal particles as antibacterial agents. Silver is one of the most well known metal particles that displays antibacterial properties. Silver sulfadiazine, for example, is an antibiotic commonly found in cream for burn wounds. Silver is usually inert, but when ionized however, it becomes highly reactive and is able to bind to proteins, destroy cells walls and membranes of bacteria.[35] The exact mechanism of this antibacterial activity for silver is still not known. However,

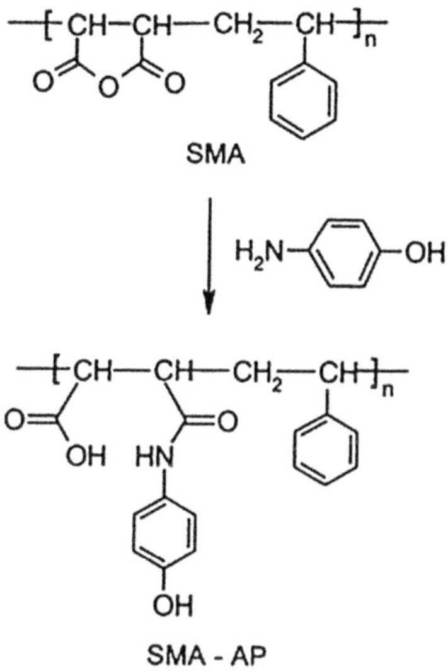

Figure 5.10 A route of chemical synthesis for SMA–AP.

results from experiments show changes in the morphology and structure of the bacterial cells before and after interactions with silver ions. Silver ions disrupt both gram-positive and gram-negative bacterial cell walls but are able to penetrate gram-negative bacteria more effectively as gram-positive bacteria have a thicker cell wall of peptidoglycans.[36] Reducing the size of the particles increases the surface area, allowing greater interaction between the silver ions and bacteria, resulting in greater antibacterial activity.[37] Polymers can be embedded, coated or mixed with silver nanoparticles to produce antibacterial properties. Silver nanoparticles can be embedded in polyvinyl alcohol (PVA), as PVA acts as both a reducing agent[38] and a matrix for the dispersed particles. Electrospun PVA nanofibers embedded with silver nanofibers tested against *S. aureus* and *Klebsiella pneumoniae* were found to have 99.9% reduction in bacterial colonies[39] and can be used for medical and surgical textiles such as wound dressings. Polyvinyl alcohol-poly(*N*-vinyl pyrrolidone) hydrogels containing silver nanoparticles also showed strong antibacterial activity when tested against *E. coli* and *S. aureus* bacterial cells[40] and can be used as membranes for burn wound dressings. Melt processing of silver particles loaded into polyamide[41] or polypropylene[42] produces a composite material that displays antibacterial activity due to the slow release of Ag^+ ions. Water molecules are also able to diffuse into the composite to accelerate this oxidation process, but a slower release of Ag would show more long-term antibacterial activity. Silver nanoparticles can also be loaded into poly(methyl methacrylate) bone cement and have shown to be effective *in vitro*,[43] however, *in vivo* studies have yet to be conducted. Silver is known to be more toxic to bacteria as compared to other metals and, at the same time, less toxic to mammalian cells (eukaryotes).[44]

However, silver nanoparticles pose a potential risk, as they can cause argyrosis or argyria, which results due to an accumulation of silver compounds in the human body over a long period of time. In addition, silver ions can be cytotoxic to cells such as fibroblasts. Currently, the long-term side effects of nanoparticles are still not yet fully understood as extensive studies of this have yet to be carried out.

5.5 Discussion and Comparison

By comparing these 3 classifications of antibacterial polymers, we can see that there are some disadvantages and advantages of each category. (1) Polymers with inherent antibacterial activity are shown to be effective alone, without any addition or modification required. However, in many examples of current research, modification of these polymers often helps to further improve the antibacterial properties or enhance other properties such as mechanical or chemical properties. (2) Chemically modified polymers require the addition of groups that possess antibacterial properties to a non-antibacterial main chain polymer, in order for the polymer to have antibacterial activity. In this case, there is a need to ensure that the other properties of the original main chain polymer do not change by very much. (3) Addition of antimicrobial agents provides the polymer with strong antibacterial activity. However, the antibacterial agents may leach out, and once depleted, the polymer then loses its antibacterial activity.

5.6 Future Perspectives

Conventional methods to combat bacteria include the use of small molecule antibacterial agents like disinfectants such as chlorine, which can leave toxic residues and only have a short-term effect. In addition, the increasing use of antibiotics to treat bacterial infection gives rise to more resistant strains and antibiotics. Hence, there is a need for further development in this area of antibacterial polymers as possible substitutes for conventional biocides and antibiotics.

Current antibacterial polymers listed in this chapter include natural polymers, water soluble/insoluble polymers and cationic polymers, all of which are currently being developed with the intention of being used for various applications. Current applications include antibacterial surfaces of medical implants, devices and biofilms with self-cleaning capabilities which kill any bacteria that adsorb onto its surface. Antibacterial coatings are also being applied onto fibers and textiles such as cotton or onto medical devices and implants to prevent bacterial infection. Antibacterial polymers can also be used as porous membranes or filters for water treatment and purification. Polymers used as antibacterial packaging minimize the use of preservatives and other additives, while preserving the quality and extending the shelf life of the products.

5.6.1 Clinical Trials

Some antibacterial polymers are already in various stages of clinical trials. Chitosan-based biomedical products,[45] such as HemCon and ChitoFlex, have already been introduced into the market. BioWeld1, which is a chitosan film for surgical incision closure of the skin, on the other hand, is still under ongoing clinical trials. Acrylate derivatives as a surface biofilm coating onto dental crowns have already completed clinical trials, while their use in dental composites as filling materials is still under clinical trials.

5.6.2 Future Research

The focus of research on new antibacterial polymers in the future would aim to not just improve the strength of the antibacterial property but also to improve its effectiveness across a wider spectrum of different bacterial strains, which includes both gram-positive and gram-negative bacteria. The long-term effectiveness and stability of the polymers in different conditions is also important. Cytotoxicity and biocompatibility are also critical properties that need to be considered, especially if the antibacterial polymers are to be used in biomedical implants or devices. Techniques are being developed to improve the surface properties of biomaterials, such as immobilizing antibacterial polymers on the surface of biomaterial, so that only the bulk properties and function of the original biomaterial are preserved. Non-leaching antibacterial coatings for implants are also being developed to ensure that the antibacterial properties of the coatings last longer and do not release any cytotoxic by-products into the human body.

References

1. S. Josset, N. Keller, M.-C. Lett, M. J. Ledoux and V. Keller, *Chem. Soc. Rev.*, 2008, **37**, 744–755.
2. L. Timofeeva and N. Kleshcheva, *Appl. Microbiol. Biotechnol.*, 2011, **89**, 475–492.
3. E.-R. Kenawy, S. D. Worley and R. Broughton, *Biomacromolecules*, 2007, **8**, 1359–1384.
4. W. Yuan, J. Wei, H. Lu, L. Fan and J. Du, *Chem. Commun.*, 2012, **48**, 6857–6859.
5. R. J. Cornell and L. G. Donaruma, *J. Polym. Sci., Part A: Gen. Pap.*, 1965, **3**, 827–828.
6. C. R. Allan and L. A. Hadwiger, *Exp. Mycol.*, 1979, **3**, 285–287.
7. Z. Zhong, R. Xing, S. Liu, L. Wang, S. Cai and P. Li, *Carbohydr. Res.*, 2008, **343**, 566–570.
8. Y. C. Chung, Y. P. Su, C. C. Chen, G. Jia, H. L. Wang, J. C. Wu and J. G. Lin, *Acta Pharmacol. Sin.*, 2004, **25**, 932–936.
9. T. Phaechamud, *J. Pharm. Sci. Technol.*, 2008, **1**, 2–9.

10. Y.-C. Chung, C.-L. Kuo and C.-C. Chen, *Bioresour. Technol.*, 2005, **96**, 1473–1482.
11. Y. Xie, X. Liu and Q. Chen, *Carbohydr. Polym.*, 2007, **69**, 142–147.
12. S. Tripathi, G. K. Mehrotra and P. K. Dutta, *Int. J. Biol. Macromol.*, 2009, **45**, 372–376.
13. K.-S. Chen, Y.-A. Ku, C.-H. Lee, H.-R. Lin, F.-H. Lin and T.-M. Chen, *Mater. Sci. Eng., C*, 2005, **25**, 472–478.
14. D. Gupta and A. Haile, *Carbohydr. Polym.*, 2007, **69**, 164–171.
15. T. Jiang, M. Deng, R. James, L. S. Nair and C. T. Laurencin, *Acta Biomater.*, 2014, **10**, 1632–1645.
16. G. Lu, D. Wu and R. Fu, *React. Funct. Polym.*, 2007, **67**, 355–366.
17. N. Kawabata and M. Nishiguchi, *Appl. Environ. Microbiol.*, 1988, **54**, 2532–2535.
18. T. Zhao and G. Sun, *J. Appl. Microbiol.*, 2008, **104**, 824–830.
19. B. Gao, X. Zhang and Y. Zhu, *J. Biomater. Sci., Polym. Ed.*, 2007, **18**, 531–544.
20. I. Yudovin-Farber, J. Golenser, N. Beyth, E. I. Weiss and A. J. Domb, *J. Nanomater.*, 2010, **2010**, 1–11.
21. C. Z. Chen, N. C. Beck-Tan, P. Dhurjati, T. K. van Dyk, R. A. LaRossa and S. L. Cooper, *Biomacromolecules*, 2000, **1**, 473–480.
22. M. K. Calabretta, A. Kumar, A. M. McDermott and C. Cai, *Biomacromolecules*, 2007, **8**, 1807–1811.
23. M. Zasloff, B. Martin and H. C. Chen, *Proc. Natl. Acad. Sci.*, 1988, **85**, 910–913.
24. F. Siedenbiedel and J. C. Tiller, *Polymers*, 2012, **4**, 46.
25. Y. Ishitsuka, L. Arnt, J. Majewski, S. Frey, M. Ratajczek, K. Kjaer, G. N. Tew and K. Y. C. Lee, *J. Am. Chem. Soc.*, 2006, **128**, 13123–13129.
26. L. Arnt, K. Nüsslein and G. N. Tew, *J. Polym. Sci., Part A-1: Polym. Chem.*, 2004, **42**, 3860–3864.
27. D. Liu, S. Choi, B. Chen, R. J. Doerksen, D. J. Clements, J. D. Winkler, M. L. Klein and W. F. DeGrado, *Angew. Chem., Int. Ed.*, 2004, **43**, 1158–1162.
28. J. Lin, C. Winkelman, S. D. Worley, R. M. Broughton and J. F. Williams, *J. Appl. Polym. Sci.*, 2001, **81**, 943–947.
29. J. Lin, C. Winkelmann, S. D. Worley, J. Kim, C. I. Wei, U. Cho, R. M. Broughton, J. I. Santiago and J. F. Williams, *J. Appl. Polym. Sci.*, 2002, **85**, 177–182.
30. M. R. Badrossamay and G. Sun, *Macromolecules*, 2009, **42**, 1948–1954.
31. J.-H. Jeong, Y.-S. Byoun and Y.-S. Lee, *React. Funct. Polym.*, 2002, **50**, 257–263.
32. H. Patel, D. A. Raval, D. Madamwar and S. R. Patel, *Die Angewandte Makromolekulare Chemie*, 1998, **263**, 25–30.
33. J. S. Patel, S. V. Patel, N. P. Talpada and H. A. Patel, *Die Angewandte Makromolekulare Chemie*, 1999, **271**, 24–27.
34. J. H. Jeong, Y. S. Byoun, S. B. Ko and Y. S. Lee, *J. Ind. Eng. Chem.*, 2001, 7, 310–315.
35. M. Rai, A. Yadav and A. Gade, *Biotechnol. Adv.*, 2009, **27**, 76–83.

36. Q. L. Feng, J. Wu, G. Q. Chen, F. Z. Cui, T. N. Kim and J. O. Kim, *J. Biomed. Mater. Res.*, 2000, **52**, 662–668.
37. J. Soukupová, L. Kvítek, A. Panáček, T. Nevěčná and R. Zbořil, *Mater. Chem. Phys.*, 2008, **111**, 77–81.
38. L. Longenberger and G. Mills, *J. Phys. Chem.*, 1995, **99**, 475–478.
39. K. H. Hong, J. L. Park, I. H. Sul, J. H. Youk and T. J. Kang, *J. Polym. Sci., Part B: Polym. Phys.*, 2006, **44**, 2468–2474.
40. H. Yu, X. Xu, X. Chen, T. Lu, P. Zhang and X. Jing, *J. Appl. Polym. Sci.*, 2007, **103**, 125–133.
41. C. Radheshkumar and H. Münstedt, *Mater. Lett.*, 2005, **59**, 1949–1953.
42. C. Radheshkumar and H. Münstedt, *React. Funct. Polym.*, 2006, **66**, 780–788.
43. V. Alt, T. Bechert, P. Steinrücke, M. Wagener, P. Seidel, E. Dingeldein, E. Domann and R. Schnettler, *Biomaterials*, 2004, **25**, 4383–4391.
44. G. Zhao and S. E. Stevens, Jr., *BioMetals*, 1998, **11**, 27–32.
45. A. Jain, L. S. Duvvuri, S. Farah, N. Beyth, A. J. Domb and W. Khan, *Adv. Healthcare Mater.*, 2014, **3**, 1969–1985.

CHAPTER 6

Four-Dimensional (4D) Printing in Consumer Applications

XIAN JUN LOH*[a,b]

[a]Institute of Materials Research and Engineering, A*STAR (Agency for Science, Technology and Research), 2 Fusionopolis Way, Innovis, #08-03, Singapore 138634, Singapore; [b]Department of Materials Science and Engineering, National University of Singapore, Singapore 117574, Singapore
*E-mail: lohxj@imre.a-star.edu.sg

6.1 A Primer on 3D Printing

Today, we can download an item's specification from the Web and customize it to our own inclination or taste, then send those data to a desktop machine which can fabricate the item for us in that instance. This rising innovation is called additive manufacturing or 3D printing. 3D printing was invented in 1983 by Charles Hull. This technology has been around for about 30 years now but the interest surrounding this technology has just recently begun to take off. The slow takeup of 3D printing can be attributed to several factors: firstly, they have been excessively wasteful, secondly, the machines have not been sufficiently quick, and thirdly, the 3D printing machines themselves have been very costly. Since the heightening of the interest in 3D printing, the technique has been used to print anything from footwear to stainless steel rings to plastic mobile phone covers to titanium spinal inserts, and metallic automotive parts.

RSC Polymer Chemistry Series No. 20
Polymers for Personal Care Products and Cosmetics
Edited by Xian Jun Loh

Published by the Royal Society of Chemistry, www.rsc.org

In customizing a structure by 3D printing, the spatial information of the desired item has to be transferred into a machine. During the manufacturing process of the item, an extrusion process occurs layer by layer according to the design file and the item will be "printed" in a manner which is similar to the printing of a word processing document using a desktop printer. 3D printing commonly peruses computer aided design (CAD) information, which requires the help of a technical professional. This information is sent to a machine that cuts it into two-dimensional cross sections of that item; the machine then processes that information, layer by layer. The item gets printed, beginning at the base of the item and storing material, layer upon layer, mixing the new layer of materials to the old layer in an additive process. This material is typically in a fluid form or a material powder structure. The production procedure can happen by either firstly softening the material followed by layering or first layering and then heating the structure to mold the shape. For example, using a laser sintering machine, one can utilize a laser to intertwine the new material layer to the old layer with high precision. The main drawback is that it usually takes several hours before a physical item can be fabricated, which is not ideal for large scale production. 3D printing is not only a consumer tool; in fact, its presence in the scientific community allows for development and utilization in new fields. The use of 3D printing has also been extended to soft materials that require intricate design. They have been applied to the design of microswimmers and bio-hybrid robotics,[1] development of scaffolds for tissue engineering,[2,3] and development of microfluidic devices.[4] Au *et al.* attempted to fabricate a valve entirely from polymer by stereolithography in order to ease the various difficulties encountered in the fabrication of polydimethylsiloxane valves, such as requiring substantial technical know-how and the need for access to expensive fabrication amenities.[5] The efficient valving of microchannels is also demonstrated. 3D printing allows valves to be easily added to the overall design as modules to form larger operational units such as multi-way switches and pumps.

Yet, with all the excitement around 3D printing, the macrostructure that is built remains inherently non-responsive. To this end, besides the ability to customize a structure to the shape and size of one's desires, there does not seem to be any other additional benefit of using 3D printing. Skylar Tibbits from MIT used a multi-material blend to create a water-responsive 3D printed structure, giving rise to a new development in "4D printing." The use of the water-responsive polymer is an example of using an *adaptive* or stimuli-responsive material to elicit some type of reaction when the structure comes into contact with an external trigger. For the materials scientist, adaptive or stimuli-responsive materials have been extensively studied, as we will show.

6.2 Soft Adaptive Materials

There are many types of adaptive materials. For the purposes of this chapter, I will focus on two types which are potentially useful for 3D printing. They are stimuli-responsive hydrogels and shape memory polymers (SMP). Of the types of stimuli-responsive hydrogels, physically cross-linked or supramolecular hydrogels bear the most promise for 3D printing because they have the potential to be remolded and the structures are not set in a permanent form after printing.[6–8] These materials have been either extensively reviewed or reported in the past and the interested reader is directed to the references.[9–13] Soft materials such as alginate, gelatin, and collagen have all been printed by 3D printing techniques.[14–16] Tough hydrogels also have potential for printing. Spinks and coworkers showed the rapid prototyping of hydrogel-based fiber composites.[17] An artificial meniscus cartilage mimicking the original 3D shape and further having fiber reinforcements was printed. This is aimed at developing 3D printed forms of soft tissues possessing a spatial difference in arrangement and features leading to biological function. 3D printed structures can also be useful for bone regeneration scaffolds. The effect of a polyester, polycaprolactone, coating alendronate release profile and local drug delivery on animal model osteogenesis from polyester-coated 3D printed porous tricalcium phosphate scaffolds was investigated, and the scaffolds can be custom fabricated to fit the implant site.[18] Furthermore, the experiments show that these scaffolds are useful for local drug delivery for enhanced osteogenesis for early wound healing. On the other hand, typical SMPs possess a permanent shape which can be molded into a secondary shape through thermal or physical manipulation. Recovery of the original shape can be facilitated by exposure of the material to temperature stimulus.[19–23] Other stimuli such as water,[24–26] light,[27,28] and electric current[29–31] have been used for shape recovery purposes. Recently, Boydston and coworkers successfully showed the printing of mechano-responsive polymers.[32] Through this work, 3D printing of mechano-responsive materials entrapped within a commercially available polymer matrix is demonstrated. This process greatly simplifies the production of materials that are challenging to make with conventional engineering methods. A prototype force sensor was also developed, showing the advantages of using the additive manufacturing method to make this structure (Figure 6.1). White and coworkers recently reported generating a light-activated SMP by using an amorphous liquid crystal polymer network material (Figure 6.2).[33] This material can be photo-fixed into a temporary state within 5 minutes using 442 nm light. Recovery of the shape can be activated through thermal or optical means. This material was thermally fixed as a catapult and later used to convert light energy into mechanical work. The authors further demonstrated the launching of an object at a rate of 0.3 m s^{-1}. Mechanically-adaptive materials made from nanocomposites can also react to an external stimulus, which affects the composite material thereby changing its bulk mechanics.[34]

6.3 (3 + 1)D = 4D: Early Promises

4D printing is a stimulating up-and-coming technology for the creation of dynamic devices which have shape changing capabilities or on-demand capabilities over time.[35–37] Through the innovative combination of smart adaptive

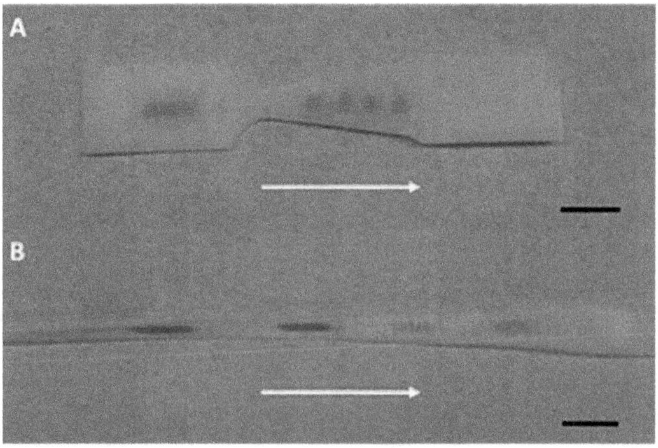

Figure 6.1 (A) 1 mm thick force sensor before elongation, and (B) post-elongation. The white arrows indicate the direction of necking. Scale bars = 10 mm.[32]

materials (described in the previous section) and additive manufacturing techniques such as 3D printing, 4D printing offers a pioneering, adaptable, and useful process for designing customized sensors, robotics, and self-assembled macrostructures.[36] Qi and coworkers have established another technique for the construction of 4D structures. Qi's work can be distinguished from Tibbits' with regards to the understanding of the molecular interactions in the materials. Additionally, they place greater emphasis on the fundamental mechanisms behind the observed phenomena. The group looked at using SMP composites embedded with glassy fibers to yield 4D structures. This system, which comprises of many different materials, is printed from a CAD file. The CAD file specifies the fiber architecture by controlling the material's anisotropic and thermo-mechanical performance. The pre-determination of the reaction of these active composite materials upon external stimulation can be achieved by controlling the quantity, site, and alignment of these fibers. During the fabrication process, the group used SMP fibers which were sandwiched between an elastomeric matrix. The spatial variation of the 3D material properties, as well as the regulation of the lamina and the laminate architecture, allows the responsive layer to self-assemble into complex 3D configurations by bending, curling, and winding, as a function of time. This aspect of the material, which changes with time, leads to the 4D aspect. This idea can be further developed by using other useful material properties as well as using shape and topology variations to achieve configuration changes.[38] In another report, Qi and coworkers showed that it is possible to print a shape memory material based on epoxy and tailor it to have a functional gradient.[39] The group demonstrated both helical and self-interlocking shape memory behavior as well as the precisely shape changing sequence utilized to reach the desired material configurations. It is easy to extend this concept and implement the polymer shape memory effect to the development of self-adjusting structures.

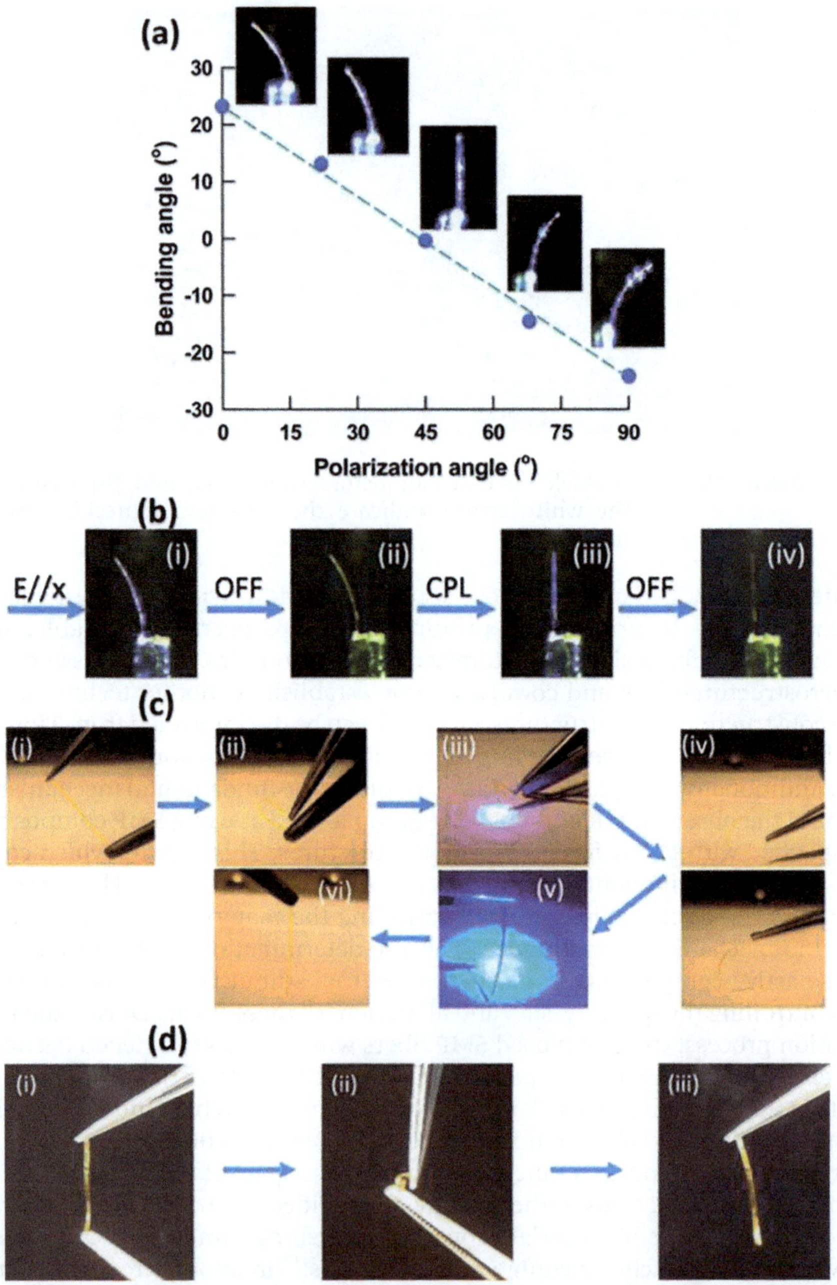

Figure 6.2 (a) The magnitude of the bending angle as a function of linear polarization angle for a cantilever composed of PD-20CL exposed to 80 mW cm^{-2} 442 nm light. (b) Shape-retention of the bent state (i) upon removal of the linearly polarized 442 nm light (ii). Exposure to right-handed circularly polarized 442 nm light (iii) unlocks the photoreconfigured state resulting in all-optical restoration of the permanent state (iv). (c) Light-activated shape memory of a freestanding PD-20CL film

Stimuli-responsive materials, which change their volume upon exposure to stimuli, can be incorporated into multi-material structures for 4D printing. One of the potential applications is in the "printing" of an artificial muscle which can mimic the action of biological muscles. However, current materials are mechanically too weak to behave as a true muscle, leading to the development of tough hydrogels as possible synthetic alternatives.[40–42] Recent reports covering 4D printing use water or temperature to demonstrate stimuli-triggered change of shape.[36,38] However, these materials respond very slowly, possess very limited reversibility, and the motion of the material is restricted to just bending motions that generate little force. Brinks *et al.* describe fast and reversible 3D printed tough hydrogel materials made from an alginate/*N*-isopropylacrylamide interpenetrating network and further demonstrate their integration into a smart valve that regulates water flow in a device. This key development brings us closer to a possible soft mechanical actuator. In another report, DNA adhesives can be used as supramolecular anchors to link up macrostructures.[43] Microparticles were assembled into a free-standing colloidal gel through extrusion with a 3D printer (Figure 6.3). The microscale structure can be manipulated by adjusting the reversible adhesion between particles. This technology is also biofriendly to cells, being able to encapsulate the cells within its matrix.

6.4 Outlook and Perspectives

This is a period for key developments in new materials that can be 3D printed. There are many ideas bustling around with the potential for programmable carbon fiber, programmable wood, and programmable textiles. These materials serve to impact potential industries such as aerospace, automotive, clothing, construction, defense and military, healthcare and utility. The total 4D printing market has a projected value of $63.0 million in 2019 and is expected to reach $555.6 million by 2025, at a compound annual growth rate of 42.98% from 2019 to 2025.[44] The excitement and interest surrounding the future of additive manufacturing is evident. For example, the US Army Research Office awarded US$855 000 worth of funds to a team comprising members from Harvard's School of Engineering and Applied Science, The University of Illinois, and The University of Pittsburgh Swanson School of Engineering to work on the new printing technologies as well as to study the fabrication of adaptive and biomimetic composites with (re)programmable shapes, properties, or

(i) permanent shape, (ii) mechanical deformation, (iii) photo-fixing, and (iv) shape retention (in the absence of light). Exposure to right-handed circularly polarized light unlocks the photo-fixed state (v) allowing for recovery of the permanent shape (vi). (d) In the absence of photo-fixing, mechanical deformation of PD-20CL does not retain the deformed state: (i) permanent shape, (ii) mechanical deformation, and (iii) restoration of permanent shape after removal of mechanical deformation. Figure reproduced from ref. 33 with permission from The Royal Society of Chemistry.

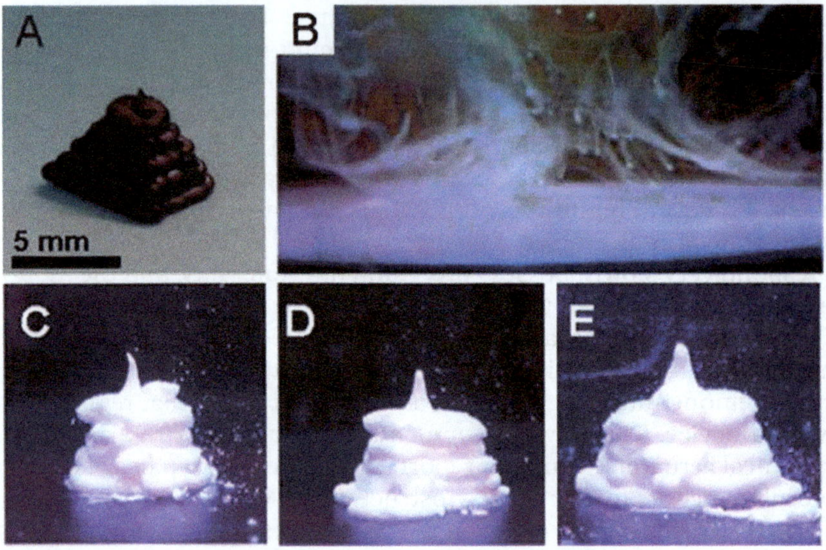

Figure 6.3 3D printing of DNA-derivatized microparticles. (A) Digital photograph shows an ABS thermoplastic pyramid printed by the 3D printer to show the desired output pattern. Scale bar is 5 mm throughout. (B) Digital micrograph shows the output of the 3D printer when it attempts to extrude microparticles bearing noncomplementary DNA. (C–E) Digital micrographs show DNA cross-linked colloidal gel printed into the pyramidal shape with extrusion rates of 1.3, 1.7, and 2.1 $\mu L\ s^{-1}$, respectively. The 3D printer head motion pattern was identical in all three cases. Figure reproduced with permission from P. B. Allen, Z. Khaing, C. E. Schmidt and A. D. Ellington, *ACS Biomater. Sci. Eng.*, 2015, **1**, 19–26. Copyright (2015) American Chemical Society.[43]

functionality upon exposure to external stimuli. In the coming years, landmark patents covering laser sintering techniques for rapid prototyping will expire and this can trigger massive development in the fabrication of open-source machines. One of the issues facing adopters of 3D printing is the speed of printing. Planar curing methods could lead to faster build times. Other methods could involve multiple light sources for the curing of different parts of the structure. Another factor which is important is the precise selection of material at a designated printing site. The development of materials which are suitable for printing is critical for the success of this technique. In fact, to utilize this technique for biomedical applications, there needs to be a concerted effort to develop machines that can print from aqueous medium. There has also got to be a portfolio of materials that are developed to fit such a printer. In this world of viral videos and social media, every new technology is looked upon with keen interest. In order to sustain this interest, the scientific community will have to continuously improve the system and the materials that are fabricated from these systems and also stay relevant to the practical needs of the global population.

References

1. M. M. Stanton, C. Trichet-Paredes and S. Sanchez, *Lab Chip*, 2015, **15**, 1634–1637.
2. J.-F. Xing, M.-L. Zheng and X.-M. Duan, *Chem. Soc. Rev.*, 2015, **44**, 5031–5039.
3. S. Zhao, M. Zhu, J. Zhang, Y. Zhang, Z. Liu, Y. Zhu and C. Zhang, *J. Mater. Chem. B*, 2014, **2**, 6106–6118.
4. K. G. Lee, K. J. Park, S. Seok, S. Shin, D. H. Kim, J. Y. Park, Y. S. Heo, S. J. Lee and T. J. Lee, *RSC Adv.*, 2014, **4**, 32876–32880.
5. A. K. Au, N. Bhattacharjee, L. F. Horowitz, T. C. Chang and A. Folch, *Lab Chip*, 2015, **15**, 1934–1941.
6. Y. Cui, M. Tan, A. Zhu and M. Guo, *J. Mater. Chem. B*, 2014, **2**, 2978–2982.
7. X. Yan, F. Wang, B. Zheng and F. Huang, *Chem. Soc. Rev.*, 2012, **41**, 6042–6065.
8. Z. Yu, J. Zhang, R. J. Coulston, R. M. Parker, F. Biedermann, X. Liu, O. A. Scherman and C. Abell, *Chem. Sci.*, 2015, **6**, 4929–4933.
9. Z. Li and X. J. Loh, *Chem. Soc. Rev.*, 2015, **44**, 2865–2879.
10. H. Ye, C. Owh and X. J. Loh, *RSC Adv.*, 2015, **5**, 48720–48728.
11. J. Y. Zheng, M. J. Tan, P. Thoniyot and X. J. Loh, *RSC Adv.*, 2015, **5**, 62314–62318.
12. E. A. Appel, J. del Barrio, X. J. Loh and O. A. Scherman, *Chem. Soc. Rev.*, 2012, **41**, 6195–6214.
13. X. J. Loh, *Mater. Horiz.*, 2014, **1**, 185–195.
14. S. E. Bakarich, M. i. h. Panhuis, S. Beirne, G. G. Wallace and G. M. Spinks, *J. Mater. Chem. B*, 2013, **1**, 4939–4946.
15. C. Colosi, M. Costantini, R. Latini, S. Ciccarelli, A. Stampella, A. Barbetta, M. Massimi, L. Conti Devirgiliis and M. Dentini, *J. Mater. Chem. B*, 2014, **2**, 6779–6791.
16. J. Y. Park, J.-H. Shim, S.-A. Choi, J. Jang, M. Kim, S. H. Lee and D.-W. Cho, *J. Mater. Chem. B*, 2015, **3**, 5415–5425.
17. S. E. Bakarich, R. Gorkin, M. I. H. Panhuis and G. M. Spinks, *ACS Appl. Mater. Interfaces*, 2014, **6**, 15998–16006.
18. S. Tarafder and S. Bose, *ACS Appl. Mater. Interfaces*, 2014, **6**, 9955–9965.
19. Y. Bai, X. Zhang, Q. Wang and T. Wang, *J. Mater. Chem. A*, 2014, **2**, 4771–4778.
20. C. Liu, H. Qin and P. T. Mather, *J. Mater. Chem.*, 2007, **17**, 1543–1558.
21. M. Ragin Ramdas, K. S. Santhosh Kumar and C. P. Reghunadhan Nair, *J. Mater. Chem. A*, 2015, **3**, 11596–11606.
22. I. V. W. Small, P. Singhal, T. S. Wilson and D. J. Maitland, *J. Mater. Chem.*, 2010, **20**, 3356–3366.
23. Y. Wu, J. Hu, J. Han, Y. Zhu, H. Huang, J. Li and B. Tang, *J. Mater. Chem. A*, 2014, **2**, 18816–18822.
24. X. Gu and P. T. Mather, *RSC Adv.*, 2013, **3**, 15783–15791.
25. X. Qi, X. Yao, S. Deng, T. Zhou and Q. Fu, *J. Mater. Chem. A*, 2014, **2**, 2240–2249.

26. L. Wang, X. Yang, H. Chen, G. Yang, T. Gong, W. Li and S. Zhou, *Polym. Chem.*, 2013, **4**, 4461–4468.
27. D. Habault, H. Zhang and Y. Zhao, *Chem. Soc. Rev.*, 2013, **42**, 7244–7256.
28. L. Yu, Q. Wang, J. Sun, C. Li, C. Zou, Z. He, Z. Wang, L. Zhou, L. Zhang and H. Yang, *J. Mater. Chem. A*, 2015, **3**, 13953–13961.
29. G. Fei, G. Li, L. Wu and H. Xia, *Soft Matter*, 2012, **8**, 5123–5126.
30. G. Fei, C. Tuinea-Bobe, D. Li, G. Li, B. Whiteside, P. Coates and H. Xia, *RSC Adv.*, 2013, **3**, 24132–24139.
31. C.-L. Huang, M.-J. He, M. Huo, L. Du, C. Zhan, C.-J. Fan, K.-K. Yang, I.-J. Chin and Y.-Z. Wang, *Polym. Chem.*, 2013, **4**, 3987–3997.
32. G. I. Peterson, M. B. Larsen, M. A. Ganter, D. W. Storti and A. J. Boydston, *ACS Appl. Mater. Interfaces*, 2015, **7**, 577–583.
33. K. M. Lee, H. Koerner, R. A. Vaia, T. J. Bunning and T. J. White, *Soft Matter*, 2011, **7**, 4318–4324.
34. L. Hsu, C. Weder and S. J. Rowan, *J. Mater. Chem.*, 2011, **21**, 2812–2822.
35. Q. Ge, C. K. Dunn, H. J. Qi and M. L. Dunn, *Smart Mater. Struct.*, 2014, **23**, 094007.
36. S. Tibbits, *Archit. Des.*, 2014, **84**, 116–121.
37. S. Tibbits, C. McKnelly, C. Olguin, D. Dikovsky and S. Hirsch, *Acadia 2014: Design Agency*, 2014, pp.539–548.
38. Q. Ge, H. J. Qi and M. L. Dunn, *Appl. Phys. Lett.*, 2013, **103**, 5.
39. K. Yu, A. Ritchie, Y. Mao, M. L. Dunn and H. J. Qi, *Procedia IUTAM*, 2015, **12**, 193–203.
40. Y. Sun, S. Liu, G. Du, G. Gao and J. Fu, *Chem. Commun.*, 2015, **51**, 8512–8515.
41. X. Zhao, *Soft Matter*, 2014, **10**, 672–687.
42. Z. W. Low, P. L. Chee, D. Kai and X. J. Loh, *RSC Adv.*, 2015, **5**, 57678–57685.
43. P. B. Allen, Z. Khaing, C. E. Schmidt and A. D. Ellington, *ACS Biomater. Sci. Eng.*, 2015, **1**, 19–26.
44. *4D Printing Market Worth $555.6 Million by 2025*, PR Newswire, New York, 2015, http://www.prnewswire.com/news-releases/4d-printing-market-worth-5556-million-by-2025-510872321.html.

CHAPTER 7

Nanoparticle Safety in Cosmetics

SU HUI TAN[b] AND XIAN JUN LOH*[a,b]

[a]Institute of Materials Research and Engineering, A*STAR (Agency for Science, Technology and Research), 2 Fusionopolis Way, Innovis, #08-03, Singapore 138634, Singapore; [b]Department of Materials Science and Engineering, National University of Singapore, Singapore 117574, Singapore
*E-mail: lohxj@imre.a-star.edu.sg

7.1 Introduction

Nanotechnology is an emerging field for many aspects of research and development, which involves dealing with atoms and molecules up to 100 nanometers (nm) in size.[1] In simpler terms, nanoparticles are approximately 80 000 times smaller than the width of a human hair.[2] These particles have distinctively dissimilar properties from their conventional counterparts because of their extremely small size and hence larger surface area.[1] The commercialization of nanotechnology is evident across a wide spectrum of applications such as drug delivery, *in vivo* imaging, electronics and even cosmetics (Figure 7.1).

The main area of nanotechnology application is worth acknowledging as it aims to benefit human well-being. The novel properties compared with bulk materials have brought nanotechnology to the limelight in cosmetics manufacturing. However, more research has to be done to investigate the adverse effects of these new nano-cosmetic products.

RSC Polymer Chemistry Series No. 20
Polymers for Personal Care Products and Cosmetics
Edited by Xian Jun Loh

Published by the Royal Society of Chemistry, www.rsc.org

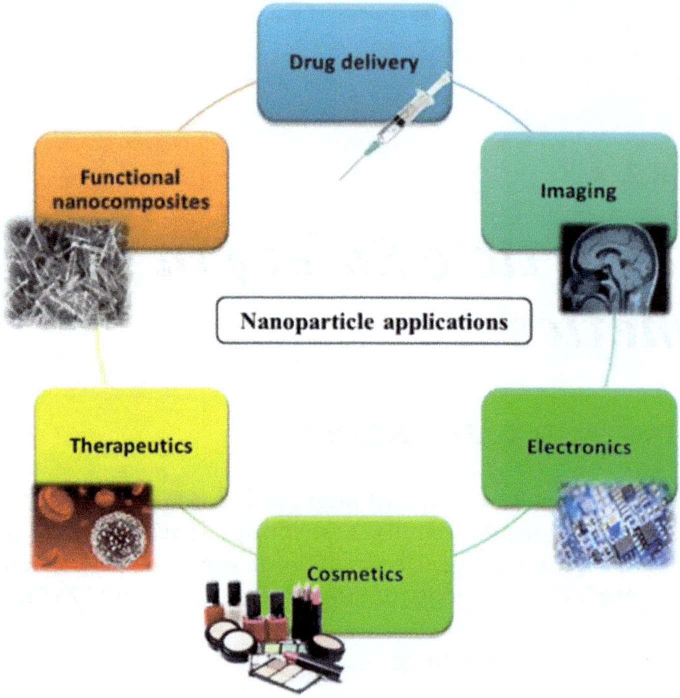

Figure 7.1 Applications of nanotechnology include drug delivery, nanoparticles such as MRI contrast agents for imaging techniques, electronics, cosmetics, therapeutics such as hyperthermia application, as well as functional nanocomposites such as reinforced plastics.

7.1.1 Overview of Cosmetics

In the USA, the Food and Drug Administration (FDA), which regulates cosmetics, defines cosmetics as "intended to be applied to the human body for cleansing, beautifying, promoting attractiveness, or altering the appearance without affecting the body's structure or functions." This broad definition includes any material intended for use as a component of a cosmetic product. The use of cosmetics dated back to as far as 10 000 BC during the ancient Egyptian times.[3] In modern times, as a result of constant innovations and changing fashion trends, a wide variety of cosmetics has been made available in the market. The availability of these inventions has undoubtedly improved the quality of life of consumers; sunscreen lotions protect the skin against harmful UV rays, soaps and mouthwashes maintain personal hygiene, while deodorants prevent body odour. According to the US FDA, cosmetics can be divided into various categories, as seen in Table 7.1. The rising affluence of consumers and changing lifestyles of people have also fuelled the global beauty care products industry and it is estimated to reach more than $250 billion by 2018.[4]

Table 7.1 Cosmetic product categories by the FDA[6]

Category	Examples
Baby products	Baby shampoos, lotions, powders, *etc.*
Bath preparations	Bath oils, bath capsules, *etc.*
Eye makeup preparations	Eyebrow pencil, eyeliner, eye lotion, *etc.*
Fragrance preparations	Cologne, perfumes, powders, *etc.*
Hair preparations (non-colouring)	Hairsprays, hair grooming aids, shampoos, *etc.*
Hair colouring preparations	Hair dyes, hair bleaches, *etc.*
Makeup preparations (not eye)	Blushers, foundations, lipsticks, body paints, *etc.*
Manicuring preparations	Basecoats, cuticle softeners, nail polish, *etc.*
Oral hygiene products	Mouthwashes, breath fresheners, *etc.*
Personal cleanliness	Bath soaps, deodorants, *etc.*
Shaving preparations	Shaving cream, shaving soap, *etc.*
Skin care preparations	Cleansing, moisturizing, masks, *etc.*
Suntan preparations	Suntan gels, creams, *etc.*

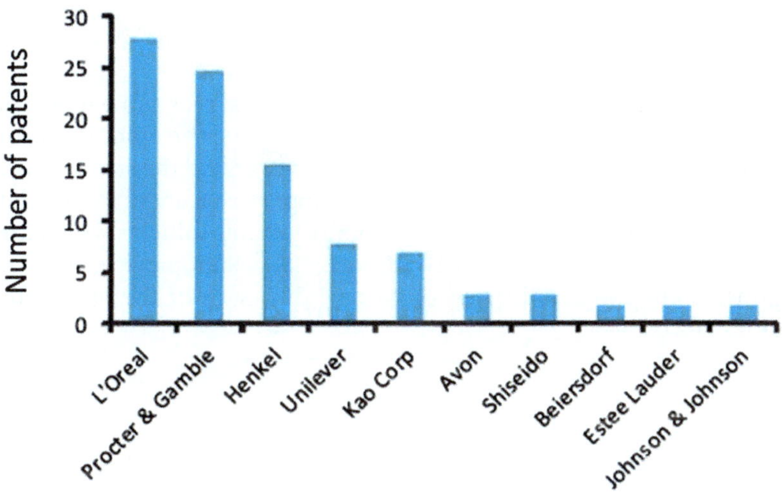

Figure 7.2 Top 10 beauty companies with the number of nano-related patents.[2] Their products can be commonly found in households all around the world.

It is not surprising that many well-known companies adopt nanotechnology in their cosmetic products. Figure 7.2 shows the ranking of the top 10 beauty companies with the number of nano-related patents. Some examples include L'Oreal's anti-wrinkle cream, Procter & Gamble's Olay brand with nanoemulsion technology, Neutrogena from Johnson & Johnson as well as products from Estee Lauder, Revlon, Lancôme and The Body Shop.[5] The use of nanotechnology in cosmetics improves penetration and allows a more effective delivery of active ingredients to the skin or hair shafts, and also enhances some properties such as colour and solubility.[2,5] While advocates of nanotechnology claim the benefits nanomaterials can bring, others raise concerns over the safety of such cosmetic products. These concerns arise because these tiny particles have different interactions with cells, which may lead to unpredicted lethal consequences. In light of the growing popularity of nanomaterials in cosmetics, it is of paramount importance to ensure their safety not only for consumers, but also for the workers handling the products as well as the environment.

7.2 Accumulation of Nanoparticles in the Body

Due to the increasing prevalence of nanotechnology, humans are constantly being exposed to nanoparticles, which can accumulate in the body. The most common methods of entry into the body are *via* skin contact, inhalation through the respiratory system as well as by ingestion. Once in the body, nanoparticles can find their way into other internal organs.

7.2.1 Skin

The skin is the largest organ of the human body and serves as a barrier against the environment. It is made up of two primary layers; the epidermis and the dermis. The epidermis has an outermost sublayer called the *stratum corneum* (SC), which is made up of keratinocytes. Nanoparticles are able to penetrate the skin barrier through the hair, sweat glands or hair follicles which are surrounded by a network of capillaries (Figure 7.3).[7] Nanoparticle penetration through the skin can also be due to passive diffusion or the application of mechanical force.[8]

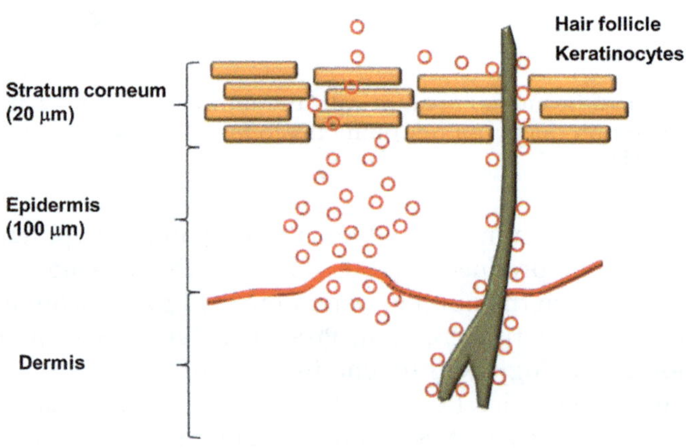

Figure 7.3 Layers of the skin and possible nanoparticle penetration pathways.[7]

7.2.2 Respiratory System

The airways, lungs and respiratory muscles make up the respiratory system. Dry nanoparticles become airborne easily and can be effortlessly taken up by the respiratory system during gas exchange. Any damage caused to this system can affect other biological systems as the respiratory system plays a vital role in maintaining metabolic and endocrine functions.[7]

7.2.3 Ingestion

The ingestion of nanomaterials can be intentional or unintentional. Intentional ingestion could be due to therapeutic or medicinal purposes. On the other hand, unintentional ingestion could be due to hand-to-mouth transfer following the handling of nanomaterials. While a large portion of the nanoparticles will be rapidly removed from the body through excretion, a minute amount may be retained in the body and translocate into other organs.[2] Furthermore, nanomaterials may not be entirely removed during wastewater treatment which, depending on community wastewater management practices, could result in its unintended consumption.

7.2.4 Internal Organs

The human body comprises multiple systems that cooperate to achieve physiological homeostasis. Although some nanoparticles may be eliminated, there are others that remain in the body for longer periods of time. Nanoparticles can be translocated to various organs through the blood stream. They can easily enter organs with high blood flow such as the liver and spleen. Although dedicated protection barriers are present to protect delicate organs like the brain, and even the foetus, nanoparticles can also have the ability to penetrate these barriers and pose risks.[7]

The translocation ability of nanoparticles is dependent on their size, shape, aggregation status and surface chemistry.[7] These nanoparticles invade the body systems and can lead to disturbance in cell signalling, impairment of cell and organ functions and even pathological disorders.[7] Figure 7.4 illustrates the schematic of translocation of nanoparticles in the human body, primarily through the respiratory tract.

7.3 Nanomaterials Used in Cosmetics and Their Safety

The nanomaterials used in cosmetics can be broadly classified into two groups: inorganic and organic. Organic nanomaterials contain carbon and hydrogen, while inorganic nanomaterials have either one or no carbon and hydrogen atoms. The safety of these nanomaterials will be discussed below.

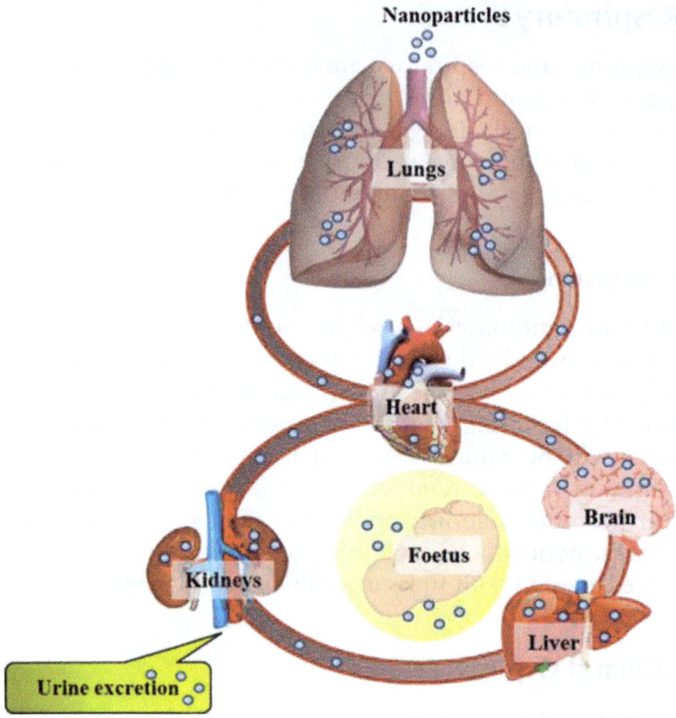

Figure 7.4 Nanoparticles circulating in different parts of the body through the blood stream, as well as clearance from the blood through urine.

7.3.1 Inorganic Nanomaterials

7.3.1.1 Titanium Dioxide and Zinc Oxide

Titanium dioxide (TiO_2) and zinc oxide (ZnO) are the main ingredients present in sunscreens, but these ingredients turn the sunscreen white and are noticeable when applied on the skin.[9] However, in nano form, they appear transparent[10] thus making them more appealing for consumers.[9] Besides this, the nano form also has the added advantage of filtering UV light more efficiently by being highly photoreactive.[10,11] These nanoparticles act as UV filters by absorption, reflection and scattering of UV photons as well as neutralizing free radicals by antioxidants.[10]

With regard to the safety of TiO_2 and ZnO nanoparticles, Filipe *et al.*'s work used three different formulations of sunscreens to explore any potential penetration of these nanoparticles. Their work concluded that the amounts of nanoparticles present in the formulations are too minimal for detection in the viable layers.[12] Another study concluded that there is a rapid reduction in TiO_2 concentration[13] with increasing depth of the SC. In other words, the nanoparticles used in sunscreens do not pose threats to human skin or health. The penetration of TiO_2 into the hair follicles was also reported.[13]

However, this phenomenon could not be regarded as penetration into the living tissues of the skin below because the follicular channel is protected by a horny layer barrier. No observation of TiO_2 penetration could be observed in the living tissues even with the use of highly sensitive methods.[13]

Darvin *et al.* reported that TiO_2 nanoparticles of more than 30 nm do not penetrate the intact skin barrier and do not enter viable layers of the epidermis. Similarly, ZnO nanoparticles of 30 nm stay on the skin surface and do not penetrate into the viable layers of the epidermis. *In vitro* tests on human and porcine skin also showed limited penetration of TiO_2 and ZnO into the upper layers of the SC and the absence of penetration into the living skin. However, if the skin is damaged by interaction with chemical irritants or mechanical procedures, various amounts of TiO_2 can be detected in the living tissue.[10] Nonetheless, there are reports indicating the inability of these particles to penetrate into or through normal and compromised human or animal skin, and that these particles are well-tolerated on the skin from oral and topical *in vivo* toxicity tests.[11]

While small amounts of UV exposure can be beneficial, prolonged exposure can be detrimental as it can degenerate skin cells, induce fibrous tissue and blood vessels. These can lead to premature skin ageing, photodermatoses, actinic keratosis and even skin cancer.[14] Although studies offer a slight difference in conclusions, it appears that TiO_2 and ZnO nanoparticles in sunscreen products generally provide more protection than risks to consumers.

7.3.1.2 Nanosilver

Nanosilver is an important ingredient in cosmetics due to its enhanced antibacterial properties.[2] Nanosilver possesses biocidal properties due to the slow release of Ag^+ ions, which have multiple mechanisms making it more difficult for bacteria to produce resistant strains.[15] Their extremely small size offers large surface areas with high surface energy and increases the number of reactive sites. Nanosilver has been incorporated into many products such as food containers, odour-resistant hygiene products, detergents, soaps and even in implantable devices and wound dressings as they all can benefit from its antibacterial properties.[16] In the case of cosmetics, nanosilver has been used in deodorants that work by killing bacteria on the skin and preventing them from causing body odour.

The toxicity of nanosilver particles is not thoroughly investigated. Inhalation experiments have been done on test animal models. Nanosilver particles were detected in the lungs and liver of rats directly after exposure, with substantial amounts found in the heart, brain and other internal organs. It was speculated that nanosilver particles entered the blood capillaries through the alveoli since the concentration in the lungs was reduced swiftly after inhalation.[17] Lankveld *et al.*'s work also reported that the concentration of nanosilver was redistributed in other organs after being injected intravenously into rats.[18] However, it was reported in another inhalation experiment that no significant health effects were found in rats after a long and high dose exposure.[19] Research also suggested that nanosilver particles had an effect on the gene expression in mouse brains, surfacing motor neuron disorders, neurodegenerative disease and immune cell functions.[16] It remains unclear whether the toxicity of nanosilver particles stemmed from the nanoparticles themselves or from the Ag^+.

The study of nanosilver particles in response to health concerns remains limited and uncertain and more research is required to allow a better understanding of the potential health hazards nanosilver particles pose to humans.

7.3.1.3 Nanogold

Nanogold also enhances antibacterial activity in products and has been incorporated into toothpastes.[2] Gold can also be used as an anti-ageing ingredient in cosmetics because of its efficiency in catalysing the formation of C–O, C–H, C=O, C–C, and N–H bonds for collagen regeneration.[20] Nanogold provides a greater surface area for catalytic ability and is transparent. It was also observed that albumin remained colourless when nanogold was added thus indicating that denaturation did not occur. This suggests that nanogold would be safe for cosmetic use on the skin.[20] Currently many cosmetic products on the market already contain nanogold formulations although this in itself is insufficient to justify the safety of nanogold.

Investigations using viability assays, gene expression analysis and cell morphology assays have been carried out to assess the toxicity of nanogold on cell cultures. It should be noted that nanogold particles are excellent light absorbers in the visible region and may hinder luminescence or fluorescence assays.[21] Several studies suggested that its toxicity could be due to other factors such as capping agents for the stabilization of nanogold particles or its degradation products, rather than the inert and non-toxic core of nanogold particles.[22,23]

In vitro permeation of gold nanoparticles conducted on rat skin revealed a size-dependent effect; an increase in size of nanoparticles results in a drop in permeation through the rat skin.[24] However, the work did not present any toxicological studies of the nanogold particles on the rat. Therefore, no conclusion could be made on whether the permeation of nanogold particles indeed presents a threat to the biological system. Although investigations on cell cultures showed toxicity of nanogold, *in vitro* results do not corroborate with *in vivo* results.

7.3.2 Organic Nanomaterials

7.3.2.1 Liposomes

Liposomes are spherical vesicles with a membrane made up of a phospholipid and cholesterol bilayer.[25] In other words, liposomes consist of hydrophobic and hydrophilic layers and are able to carry both hydrophilic and hydrophobic molecules. They are used for active ingredient delivery as the

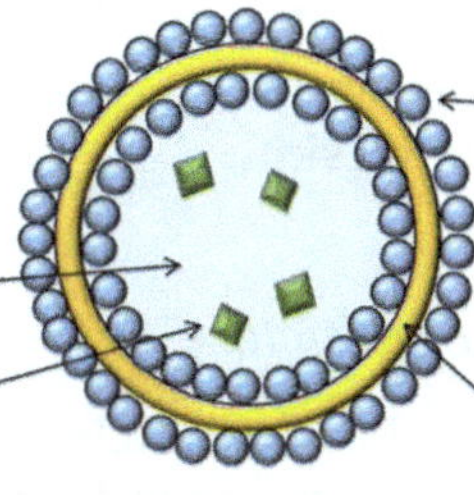

Figure 7.5 Schematic diagram of a liposome.

lipid bilayer is able to merge with the cell membrane and release the content into the cellular environment.[9,26] Liposomes are used for skin care applications as they allow better skin penetration, and help fix active ingredients to the outmost skin layers.[25] Likewise, hair loss products make use of liposomes as they allow penetration even into the deeper shafts of the hair.[9] Figure 7.5 shows a schematic representation of a liposome containing active ingredients to be delivered to the skin.

Due to the similarity between the liposome and lipids on the skin, liposomes are able to enhance skin hydration in dry skin conditions.[25] This serves as an advantage for liposomes in cosmetics applications, especially for dermatological functions. Many cosmetic ingredients make use of liposomal formulations because liposomes are non-toxic and non-invasive, and are able to deliver hydrophilic and/or lipophilic substances.[27] Furthermore, liposomal formulations have more efficient dose/effect ratios and trigger fewer reactions compared to the free substances at the same concentration.

7.3.2.2 Nanocapsules

Nanocapsules are nano-sized capsules or polymeric nanoparticles, designed to encapsulate drugs or cosmetic ingredients. Besides providing the structural support, nanocapsules are also able to have biological activity or effects depending on the content. They are able to entrap both aqueous and oily liquid cores.[9] This can include sunscreen, antibacterial and antioxidant substances. Polymeric nanocapsules have significant advantages because they are able to provide drug stability as well as control the release rate and penetration of drugs and active ingredients into the skin.[28] The polymeric shells of nanocapsules play a pivotal role as a barrier to diffusion; the rate of drug delivery is determined by the crystallinity of the polymer. The polymeric wall also serves to protect active ingredients within it from photo-degradation caused by exposure to UV radiation.

Nanocapsules restrict skin penetration of UV filters, thereby prolonging the protective effect by increasing adhesion to the SC of the skin.[2,9,28] Nanocapsules with positively-charged surfaces were proved to deliver a greater amount of active ingredients to the SC due to better adhesiveness. In addition, nanocapsules are able to modify the rheology of semisolid formulations.

A higher viscosity would in turn allow easier usage as compared to liquid suspensions. However, a higher viscosity is associated with a slower release rate of active ingredients into the skin.

7.3.2.3 Solid Lipid Nanoparticles

At body temperature, solid lipid nanoparticles are lipid particles in the solid state and are stabilized by surfactants in order to prevent them from degrading.[2,9] Solid lipid nanoparticles are formed by a matrix of biodegradable and biocompatible raw materials. In cosmetics, they are used as topical carrier systems for sustained delivery. These systems are able to protect substances that are easily broken down from chemical degradation and regulate the release of substances because of the solid lipid matrix.[27] Additionally, solid lipid nanoparticles allow the formation of films over the skin, increasing its moisture level. Furthermore, it was reported that due to the small size and high surface area of solid lipid nanoparticles, they increase the contact of active ingredients with the SC of the skin.[29,30] Consequently, there will be high permeation of the carried active ingredients through the viable skin.

Despite the advantages that solid lipid nanoparticles possess, disadvantages are certainly present. Other than having a relatively low loading capacity for certain active ingredients, there is a lack of physical stability which can cause expulsion of their loads during storage or administration.[27,28] Nevertheless, alternative carrier systems such as micelles and liposomes are available.

7.3.2.4 Nanoemulsions

Nanoemulsions are 50 to 100 nm sized droplets dispersed in an external phase, thereby forming an emulsion.[2] The small droplets can flow easily and have a pleasant touch. Nanoemulsions are able to enhance the solubility of lipophilic active ingredients and alter the transport properties due to large interfacial areas.[28] In cosmetics applications, nanoemulsions are used in moisturizers and hydrating lotions, and hair conditioners. They are even used in topical medicines for dermatological disease treatments. Nanoemulsions are used to provide controlled delivery of active ingredients to specific layers of the skin.[28] They have several advantages over suspensions and conventional emulsions such as low cost of manufacturing and long shelf life due to higher stability.

Calderilla-Fajardo *et al.*'s work showed that the incorporation of substances into nanoemulsions resulted in a higher penetration rate when compared to those in nanocapsules or conventional emulsions.[31] It was suggested that this could possibly be due to the smaller size and higher flexibility of the droplets, in comparison to larger and rigid nanocapsules and the much bigger droplet size of a conventional emulsion. Besides flexibility, nanoemulsions have a competitive edge because of their higher affinity with the SC layer of the skin compared to other polymers.[28]

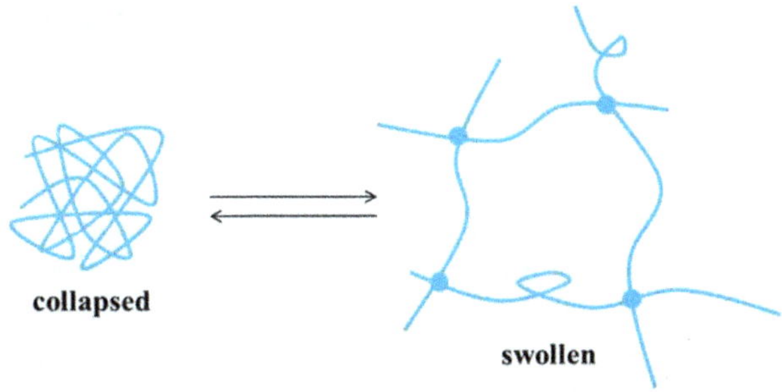

Figure 7.6 Schematic illustration of the collapse and swelling of a hydrogel in response to external stimuli.

7.3.2.5 *Hydrogels*

Hydrogels are three-dimensional polymer networks that are able to swell in fluids without dissolving themselves, due to the presence of chemical or physical crosslinks. Hydrogels can be modified to be responsive to external stimuli such as temperature, pH and solvents. The effect of external stimuli causes hydrogels to either collapse or swell (Figure 7.6).

Lee and Kim's work investigated the efficiency of pH-sensitive hydrogels in delivering cosmetic ingredients. The delivery system operates by incorporating the cosmetic ingredients into the pH-sensitive hydrogel, which is stored at a pH below the pK_a of the hydrogel.[32] At low pH, the hydrogel network is collapsed and thus unable to release the ingredients. The delivery of the ingredients occurs after application on the human skin, which is maintained at pH 6 due to human homeostasis. Eventually, the increase in pH stimulates the release of substances for absorption through the skin due to the pH-responsive swelling of the hydrogel.

The use of hydrogels in cosmetics is particularly useful as hydrogels have the potential to be used as smart carriers which can be stimulated by the external environment for cosmetic applications. However, the loading efficiency and release behaviour of the hydrogel may vary with different ingredients. More investigations are required to obtain the most efficient delivery system in cosmetics.

7.4 Safety Standards for Cosmetics

Safety standards are put in place in order to ensure that manufacturers and importers of cosmetic products abide by them for the welfare of the consumers, workers and the environment. Elaborated below are some facts regarding the safety standards implemented in different countries.

7.4.1 Cosmetic Ingredient Review (CIR)[33]

Established in 1976 with the support of the US FDA and Consumer Federation of America, the Cosmetic Ingredient Review reviews and assesses the safety of ingredients used in cosmetic products and publishes the results in peer-reviewed scientific literature. However, the CIR only reviews ingredients and not products. Fragrances, colours and flavourings are also not commonly reviewed by the CIR. The availability of all scientific literature will be made known to the public.

A priority list of ingredients to be reviewed will be developed annually, consisting of up-to-date commercially distributed cosmetic products. The annual priority list is drafted based on the frequency of use of the ingredient in different products, which is determined by the FDA's Voluntary Cosmetic Registration Program (VCRP). The prioritisation of ingredients may be further arranged based on toxicological considerations, and similar ingredients will be classified together whenever appropriate.

7.4.2 Food and Drug Administration (FDA)[34]

In the United States, cosmetic products, other than colour additives, do not require FDA premarket approval. However, enforcement action can be taken against products on the market that do not comply with the law. In other words, if the ingredients a manufacturer is using are not colour additives or ingredients that are prohibited by regulation, he will be able to use any ingredient in formulating a product as long as it is safe for specific usage, properly labelled and does not affect the quality of the cosmetic.

As the law and FDA regulations do not require specific tests to validate the safety of ingredients and products, manufacturers have a legal duty to ensure the safety of their products. Cosmetic companies are also not required to share their safety information with the FDA. However, the FDA makes efforts to constantly advise manufacturers to use sufficient and appropriate testing to ensure the safety of their products and ingredients. The safety can be substantiated through already available toxicological test data on individual ingredients and on product formulations with similar compositions, and through additional toxicological and other tests. Other than restricting the use of certain ingredients, labels on certain types of cosmetics are also necessary. The FDA does not have the authority to order recalls of cosmetics. Despite this, they do monitor companies that conduct a product recall. The FDA also inspects cosmetic facilities and import infections and addresses consumers' complaints by collecting samples for examination and analysis. Conducting research on cosmetic products and ingredients may also be done to address safety concerns.

7.4.3 Health Science Authority in Singapore[35]

Singapore's Health Science Authority (HSA) implemented the Health Products Act for the regulatory control of cosmetic products. This act requires companies or individuals who plan to sell or supply a cosmetic product in Singapore to inform the HSA before bringing the product into the local market. These products should also comply with the requirements under the ASEAN Cosmetic Directive and it is compulsory for all member countries to adhere to it.

The ASEAN Cosmetic Directive includes a list of substances which must be prohibited from the composition of cosmetic products and those that are subjected to terms and conditions. It also includes the list of substances temporarily allowed, authorised colouring agents, preservatives and UV filters.

Under the ASEAN Guidelines for safety evaluation of cosmetic products, it is necessary to consider possible interactions between ingredients with regard to potential safety. The influence on skin penetration should also be considered, especially sensitisation and systemic risks. This can be assessed using *in vitro* methods. Suitable vehicles should also be used to identify the allergic potential of ingredients. Ingredients and their usage that are not covered by current legislation should be avoided.

The toxicological profile of a raw material is obtained by analysing available data, such as the results of *in vitro*, *in vivo* and clinical testing, as well as epidemiological studies. More attention is to be paid to new ingredients or ingredients of a novel application. Suppliers are the main sources of toxicological data on ingredients and these raw materials have to comply with national legislation such as occupational safety, transport, packaging and labelling. The manufacturers of cosmetic products should initiate the collection of toxicological data and other relevant information from the suppliers. Toxicological data may be obtained from scientific literature, databases, reports issued by scientific bodies or committees, safety data sheets, personal experience with the ingredient or cosmetic products handled, as well as expert opinions. Additionally, other parameters that have to be considered include the class which the ingredient used belongs to, method of application of the cosmetic product, amount of ingredient used in the product, amount of product used per application, frequency of application, area and the type of site of skin contact, duration of contact, foreseeable misuse, the type and number of consumers and the probable amount that will enter the body.

7.4.4 National Industrial Chemicals Notification and Assessment Scheme (NICNAS) in Australia[36]

The Australian Government ensures the safety of products for consumers, workers and the environment by assessing the risks associated with the cosmetic products and ingredients manufactured or imported into Australia. A cosmetic product is a substance intended to be placed in contact with any external part of the body, which serves the purpose of protecting, altering the odour, cleansing or changing the appearance, or maintaining the good condition of the body. Additionally, the ingredients should not be prohibited and should not be a therapeutic good. The product should fall under these six cosmetic product categories; sunscreens (foundations, lip, and skin moisturizers), sunbathing products, antibacterial skin products, anti-acne products, anti-dandruff products and oral hygiene products. If they comply with the Cosmetics Standard 2007, they are being regulated by the NICNAS. Following which, they will need to meet the NICNAS requirements and labelling requirements. The authorities in Australia have a list of prohibited or restricted cosmetic chemicals which is frequently updated. Cosmetics manufacturers and importers should check the websites regularly to ensure they have followed the regulations. CIR follows a set of procedures during the review of ingredients.

7.5 Drivers of Nanomaterials Usage

With the increasing demands of consumers to have more effective cosmetics products to enhance their appearance, more active ingredients are being introduced into cosmetics. However, these active ingredients are only useful if they are successfully delivered into the living tissues to rejuvenate, cleanse and protect the skin. Nanotechnology allows deeper penetration into the skin as the small particles can pass through biological membranes with ease and enables the active ingredients to be applied and absorbed more effectively.

In addition, the usage of nanomaterials such as nanocapsules, liposomes and nanoemulsions provides an advantage of better sustained release.[9] Generally, active ingredients are highly unstable to oxidation, heat and moisture.[32] Otherwise, they are prone to decompose into biologically inactive substances. The usage of nanomaterials to protect them from the harsh surrounding conditions allows a product to last longer when in use and is therefore more beneficial to consumers.

7.6 Challenges of Nanomaterials Usage

While there are advocators of nanotechnology, the usage of nanomaterials still faces challenges. This could come from objections from consumers and workers due to safety concerns, as well as their effects on the environment. Size, chemical composition, solubility, and aggregation are some of the attributes that affect the toxicity of nanomaterials.

7.6.1 Consumers

As nanotechnology is a relatively new and emerging field of research, the raising of societal concerns is inevitable. Besides, as nanomaterial application in consumer products such as cosmetics is increasing, the exposure of consumers to these materials is escalating. However, it is evident from the booming market of beauty care products that concerns from consumers have yet to reach a point where they affect the sales of the products. Consumers should understand that the health risks that nanoparticles pose to them are dependent on the route and degree of exposure.

7.6.2 Workplace

Prolonged exposure to nanoparticles happens in the workplace when they are handled during research, production as well as testing phases. Nanoparticle exposure can occur through skin contact and through the respiratory tract.

Studies have shown that these nanoparticles can be translocated and taken up by the body. Unfortunately, the effect of nanoparticles on organs is not fully understood. Moreover, it is also difficult to track the subtle amounts of nanoparticles present in the exposed workers as highly sensitive instruments are required.

7.6.3 Environment

Through air, water and soil, nanoparticles can be released into the environment during their manufacturing, usage and disposal. For example, the release of antibacterial compounds could contaminate the water for usage and also obstruct the role of beneficial bacteria present in sewage and waste water treatment plants.[2] An article published by Scientific American reported that the presence of TiO_2 nanoparticles in sewage impedes the roles of beneficial bacteria in treating waste water.[37]

Logically, performing risk assessments for nanoparticles provides a solution to appropriate ways to eliminate side effects caused by the use of nanotechnology to the environment. However, insufficient data on toxicity and exposure deters the ability to perform such assessments. Even if present, data do not show severe threats to the environment; a substantial increase in manufacturing and use of nanoparticles in the future may alter the predictions. Additionally, it should be noted that the release of nanoparticles into the environment is not strictly confined to cosmetics applications; other applications illustrated in Figure 7.1 also increase the exposure of the environment to nanoparticles.

7.7 Discussion

7.7.1 Is Use of Nanoparticles Necessary?

The shift of technology's focus to the nano-scale has undisputedly brought about many benefits and challenges. Nanotechnology allows the creation of materials with specific properties. Stronger, lighter, increased durability, better reactivity; these are some of the traits that nanotechnology offers. As mentioned previously, nanoparticles used in cosmetics can provide better penetration and absorption, aesthetic enhancement of the product, greater clarity, and antibacterial properties. The use of nanoparticles is therefore necessary to improve the quality of lives of people.

7.7.2 Is Concern Over Its Use Necessary?

Apprehension over the use of nanoparticles is unavoidable; after all, the birth of this technology is a recent phenomenon and there is a lack of awareness among consumers. Toxicity tests of these nanomaterials in cosmetics give inconclusive or even opposing results. Furthermore, as technology continues to progress and produce even smaller-scaled materials, more uncertainties will be present. Concerns over the use of nanoparticles are therefore necessary. Based on the review above, organic nanomaterials appear to be a better choice for cosmetic ingredients compared to inorganic ones. While it may be too myopic to provide this conclusion, the wide biomedical applications of organic materials are evident to substantiate the accuracy of the statement.

7.8 Outlook

The question on whether nanoparticles are indeed safe to use in cosmetic applications remains highly debatable. Cosmetics serve to improve the quality of life for its consumers; there should be a balance between beneficial and adverse health effects of engineered nanoparticles. Currently, the available information on toxicity and accumulation of nanoparticles in living organisms is unable to provide a conclusion. In order to prevent irremediable effects on health and the environment, more studies are necessary before exploring other applications. Safety standards should also be implemented to avoid possible negative consequences. This, regrettably, is easier said than done because the use and development of nanomaterials are rising. Safety standards will need to keep up with technology in order to stay relevant and provide the best outcome for consumers, workers and the environment.

References

1. R. J. Aitken, M. Q. Chaudhry, A. B. A. Boxall and M. Hull, *Occup. Med.*, 2006, **56**, 300–306.
2. S. Raj, S. Jose, U. S. Sumod and M. Sabitha, *J. Pharm. BioAllied Sci.*, 2012, **4**, 186–193.
3. Cosmeticsinfo.org, *A History of Cosmetics from Ancient Times*, 2015, http://cosmeticsinfo.org/Ancient-history-cosmetics, accessed 11 February.
4. M. Yeomans, *Global beauty market to reach $265 billion in 2017 due to an increase in GDP*, 2015, http://www.cosmeticsdesign.com/Market-Trends/Global-beauty-market-to-reach-265-billion-in-2017-due-to-an-increase-in-GDP, accessed 28 January.
5. Nanowerk, *Nanotechnology in cosmetics*, 2015, http://www.nanowerk.com/nanotechnology-in-cosmetics.php, accessed 6 February.
6. U. S. F. a. D. Administration, *Cosmetic Product Category Codes*, 2015, http://www.fda.gov/Cosmetics/RegistrationProgram/PaperRegistration/ucm111279.htm, accessed 2 February.
7. Y. Zhang, Y. Bai, J. Jia, N. Gao, Y. Li, R. Zhang, G. Jiang and B. Yan, *Chem. Soc. Rev.*, 2014, **43**, 3762–3809.
8. J. G. Rouse, J. Yang, J. P. Ryman-Rasmussen, A. R. Barron and N. A. Monteiro-Riviere, *Nano Lett.*, 2007, **7**, 155–160.
9. F. Sullivan, *Futuretech Alert*, 2013.

10. M. E. Darvin, K. König, M. Kellner-Hoefer, H. G. Breunig, W. Werncke, M. C. Meinke, A. Patzelt, W. Sterry and J. Lademann, *Skin Pharmacol. Physiol.*, 2012, **25**, 219–226.
11. G. J. Nohynek, E. Antignac, T. Re and H. Toutain, *Toxicol. Appl. Pharmacol.*, 2010, **243**, 239–259.
12. P. Filipe, J. N. Silva, R. Silva, J. L. Cirne de Castro, M. Marques Gomes, L. C. Alves, R. Santus and T. Pinheiro, *Skin Pharmacol. Physiol.*, 2009, **22**, 266–275.
13. J. Lademann, H. J. Weigmann, C. Rickmeyer, H. Barthelmes, H. Schaefer, G. Mueller and W. Sterry, *Skin Pharmacol. Physiol.*, 1999, **12**, 247–256.
14. WHO, *Health effects of UV radiation*, 2015, http://www.who.int/uv/health/en/, accessed 27 February.
15. S. N. Luoma, *Woodrow Wilson International Centre for Scholars*, 2008.
16. S.-j. Yu, Y.-g. Yin and J.-f. Liu, *Environ. Sci.: Processes Impacts*, 2013, **15**, 78–92.
17. S. Takenaka, E. Karg, C. Roth, H. Schulz, A. Ziesenis, U. Heinzmann, P. Schramel and J. Heyder, *Environ. Health Perspect.*, 2001, **109**, 547–551.
18. D. P. K. Lankveld, A. G. Oomen, P. Krystek, A. Neigh, A. Troost – de Jong, C. W. Noorlander, J. C. H. Van Eijkeren, R. E. Geertsma and W. H. De Jong, *Biomaterials*, 2010, **31**, 8350–8361.
19. J. H. Ji, J. H. Jung, S. S. Kim, J.-U. Yoon, J. D. Park, B. S. Choi, Y. H. Chung, I. H. Kwon, J. Jeong, B. S. Han, J. H. Shin, J. H. Sung, K. S. Song and I. J. Yu, *Inhalation Toxicol.*, 2007, **19**, 857–871.
20. T. Taufikurohmah, I. G. M. Sanjaya and A. Syahrani, *J. Mater. Sci. Eng.*, 2011, **1**, 857–864.
21. E. C. Dreaden, A. M. Alkilany, X. Huang, C. J. Murphy and M. A. El-Sayed, *Chem. Soc. Rev.*, 2012, **41**, 2740–2779.
22. A. M. Alkilany, P. K. Nagaria, C. R. Hexel, T. J. Shaw, C. J. Murphy and M. D. Wyatt, *Small*, 2009, **5**, 701–708.
23. T. Niidome, M. Yamagata, Y. Okamoto, Y. Akiyama, H. Takahashi, T. Kawano, Y. Katayama and Y. Niidome, *J. Controlled Release*, 2006, **114**, 343–347.
24. G. Sonavane, K. Tomoda, A. Sano, H. Ohshima, H. Terada and K. Makino, *Colloids Surf., B*, 2008, **65**, 1–10.
25. Y. Rahimpour and H. Hamishehkar, *Expert Opin. Drug Delivery*, 2012, **9**, 443–455.
26. I. P. Kaur, M. Kapila and R. Agrawal, *Ageing Res. Rev.*, 2007, **6**, 271–288.
27. S. S. Guterres, M. P. Alves and A. R. Pohlmann, *Drug Target Insights*, 2007, **2**, 147–157.
28. F. Poletto, R. R. Beck, S. Guterres and A. Pohlmann, in *Nanocosmetics and Nanomedicines*, ed. R. Beck, S. Guterres and A. Pohlmann, Springer Berlin Heidelberg, 2011, pp. 49–68.
29. V. Jenning, M. Schäfer-Korting and S. Gohla, *J. Controlled Release*, 2000, **66**, 115–126.
30. C. S. Maia, W. Mehnert and M. Schäfer-Korting, *Int. J. Pharm.*, 2000, **196**, 165–167.

31. S. B. Calderilla-Fajardo, J. Cázares-Delgadillo, R. Villalobos-García, D. Quintanar-Guerrero, A. Ganem-Quintanar and R. Robles, *Drug Dev. Ind. Pharm.*, 2006, **32**, 107–113.
32. E. Lee and B. Kim, *Korean J. Chem. Eng.*, 2011, **28**, 1347–1350.
33. C. I. Review, 2015, http://www.cir-safety.org/, accessed 19 March.
34. U. S. F. a. D. Administration, *Cosmetics*, 2015, http://www.fda.gov/Cosmetics/, accessed 19 March.
35. H. S. Authority, *Cosmetic Products Overview*, 2015, accessed 19 March.
36. N. I. C. N. a. A. Scheme, *Cosmetics*, 2015, http://www.nicnas.gov.au/chemical-information/cosmetics, accessed 19 March.
37. M. C. E. H. News, *Scientific American*, 2009.

CHAPTER 8

Silicones: The Future for Beauty and Everyday Care

HAO XUN KUAH[b] AND XIAN JUN LOH*[a,b]

[a]Institute of Materials Research and Engineering, A*STAR (Agency for Science, Technology and Research), 2 Fusionopolis Way, Innovis, #08-03, Singapore 138634, Singapore; [b]Department of Materials Science and Engineering, National University of Singapore, Singapore 117574, Singapore
*E-mail: lohxj@imre.a-star.edu.sg

8.1 Introduction

As we look at the vast selection of products in the world today, the term "silicone" should not be too foreign as an ingredient. Silicone has been a highly successful ingredient for many products since the 1990s. Its use continues to proliferate till this day due to its superior properties and excellent versatility. Silicones are used extensively in many different fields to enhance our quality of life, ranging from silicone grease in the automotive field to household products such as cookware attributing to its non-toxic characteristics.[1] Silicones have also played a crucial role in advancements in the field of electronics and computer technologies. Due to its protective characteristics, silicones were successfully incorporated in these fields and further led to new innovations that were not possible previously.

The term silicone designates materials based on the organosiloxane group. Originating from mineral quartz, or silica, silicones may be fabricated through various reactions such as condensation and hydrolysis. Silicones are

RSC Polymer Chemistry Series No. 20
Polymers for Personal Care Products and Cosmetics
Edited by Xian Jun Loh

Published by the Royal Society of Chemistry, www.rsc.org

a large group of polymeric synthetic materials which comprise of a inorganic backbone with alternating silicon and oxygen atoms (Figure 8.1) with the silicon atom attached to organic groups.[2] Organic side groups may help to bond the inorganic backbone together. Like many other polymers, by altering the chain length, side groups and amount of cross-linking, we may vary the properties to produce an array of silicone compounds to suit the needs of various industries. Silicone compounds could exist as liquid, gel, rubbery and even in plastic form, depending on the structure and chemical makeup. This versatility has brought about much attention in many industries to date. Silicone is frequently known as polydimethylsiloxane (PDMS) or dimethicone despite having numerous different types of functional silicones in the market today.

As already mentioned, silicones have a part to play in many industries and technologies. This is no exception in the personal care industries. The initial use of personal care products dates back to centuries ago. As seen in Figure 8.2, there is a large pool of silicone products invented since as early as the 1950s. Over the past decades, there has also been an increasing use of personal care products due to increasing awareness of beauty and personal well-being. Beyond basic hand lotions, shampoos and body foam, cosmetics and protective creams are becoming increasingly popular. Moreover, expectations from consumers continue to increase as they become more conscious of the benefits of personal care. Being in a consumer-driven industry, many personal care companies are required to meet the varying needs of the consumers such as the skin feel, the aesthetical aspects or the product benefits. Silicones have become one major ingredient of investigation for these personal care products.

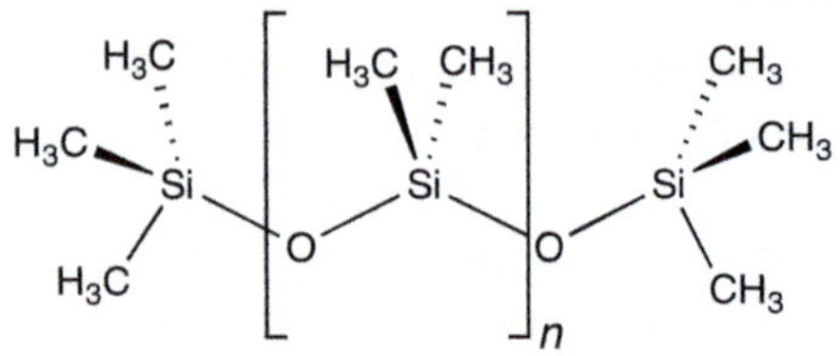

Figure 8.1 Polydimethysiloxane chemical structure.[3]

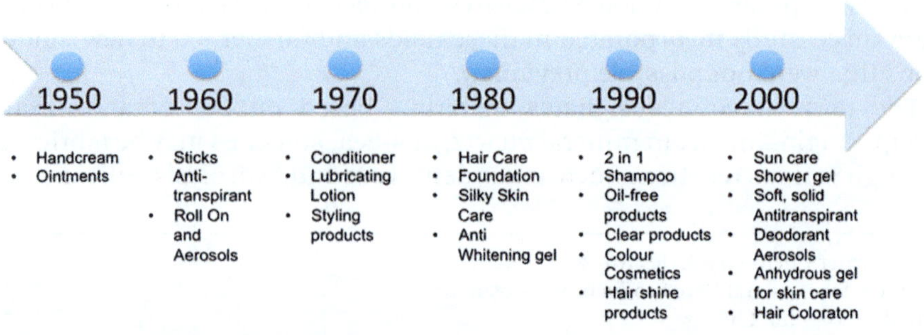

Figure 8.2 History of silicone uses in personal care.[4]

First introduced in the 1950s, upon discovering the unique characteristics of silicone, many researchers from cosmetics companies investigated the benefits of the polymer to dry, damaged skin. This subsequently led to the discovery of the first commercial personal care product that incorporated silicones. Silicone growth in personal care applications continues to grow and extend beyond various categories of personal care products. With the discovery of low molecular weight cyclosiloxanes and their volatile characteristics, silicone made another significant breakthrough in the 1970s by introducing antiperspirant actives that are non-staining and provide excellent skin feel. Silicone further made its way into hair products and cosmetics. Hair products such as styling mousses and hair conditioners included ingredients such as amino-functional polymers. By 1985, approximately 29% of personal care products included silicone in their ingredients. This percentage has doubled and, at this date, it continues to increase with a more extensive range of the silicone polymer catering to different segments of personal care products.[4]

From the chemical structure of the polymer, the siloxane backbone provides mobility and flexibility. Thus, silicones could exist in orientations where minimal energy is required for the polymer to be absorbed on a surface. The organic substitution may also spread out at interfaces, thus creating a low surface energy layer. This layer reduces intermolecular interactions, hence offering unique surface properties. PDMSs are often more chemically inert and thermally stable than other polymers due to the silicon–oxygen and silicon–carbon bonds in the polymer that are chemically stable.

In essence, silicones utilised in personal care products consist of varying types, such as linear, cyclic or organo-functional PDMS. Some may even include silicone elastomer dispersions and resins. Silicones may provide varying characteristics and function as water shields, emulsifiers or emollients in the respective products.[2] This class of polymer also provides features such as excellent spreading, high volatility, film forming abilities and permeability. Its unique attribute of being smooth, with a silky and non-greasy texture attracted many consumers to include silicones in their daily lives. Most importantly, silicones are non-toxic, which is especially vital for personal care applications.

8.2 Applications of Silicones

Personal care products may be categorised into 3 main segments: hair products, skin products, decorative cosmetics. Under the umbrella of each segment, there are different products that require varied attributes. These attributes may be extracted from the respective silicone technologies that offer solutions to the problems of current existing products.

8.2.1 Silicone and Hair Care

Human hair is a biological keratin nanocomposite fibre. Consumers' hair today is increasingly exposed to external agents such as ultraviolet (UV) rays from the sun, excessive combing and drying and chemical products, which include colour dyes. Consequently, hair products that merely cleanse are no longer sufficient. Formulators today need to deliver specialised hair products that can cleanse the hair and also repair and condition the fibre to reduce plausible hair damage.

Silicones used in typical hair care products are long, flexible siloxane backbone linkages with varying organic pendant groups bonded to the central silicon atom. Most of these products are in the liquid phase at room temperature and insoluble in water. However, addition of other functional groups such as ethylene glycol may render them water-soluble.[5] Hence, silicones could alter their chemical structure to suit different needs.

Various silicone polymers have properties that enable them to adsorb onto the negatively charged surface of the hair. Beyond that, silicone hair products often exhibit good spreadability to form a smooth film. This improves slip across and amongst hair strands, essentially helping to reduce the combing force required. Consequently, silicone hair products have good conditioning and detangling properties. Cyclomethicone, which is widely known to be a volatile cyclic silicone, is often used to formulate products with temporary conditioning. Other common silicones used in hair products include dimethicone, dimethiconol, amodimethicone and various copolymers.

Dimethicones are amongst the top silicones used in hair care. They are the core active formulating ingredients of numerous two-in-one shampoos and hair conditioners. A wide range of dimethicones is available in the market today, varying primarily in viscosities and molecular weight. Dimethicones with viscosities lesser than 5 cSt (capillary suction time) are classified to be volatile. Personal care products usually fall between 350 cSt to 12 500 cSt. Research has shown that with higher viscosity level, the conditioning effects become more significant. However, it is more difficult to formulate products with higher viscosities.[5]

To counter this problem, new formulations are investigated to increase conditioning to the hair. Gum blends such as a combination of dimethicone and dimethiconol or cyclomethicone and dimethiconol are introduced. The high molecular weight silicone gum is dispersed in relatively low molecular weight cyclomethicone or dimethicone. This gum blend functions in a two-step process. For instance, in the conditioner containing cyclomethicone and dimethiconol, cyclomethicone being volatile acts as a temporary conditioner. With its gradual evaporation, it enhances the wet combing properties of the hair. Conversely, the non-volatile dimethiconol functions as a permanent conditioner. It remains on the hair for a longer period of time and subsequently helps to improve dry combing and texture of the hair.

Amino-functional silicones are also widely used in hair products for their cationic capabilities to enhance deposition on our negatively charged hair. Commercial conditioner generally adsorbs on the hair surface through van der Waals attractions, which may be easily removed from the contact region. Amino silicones adsorb to the hair through strong electrostatic binding between the polar amino and the hair surface. Hence, they provide improved load-bearing properties and may function as a cushion in reducing hair surface damage. To further improve conditioning properties, a patented silicone resin known as aminopropyl phenyl trimethicone, with an additional phenyl group was formulated.[6] The phenyl group has a high refractive index, which aids in imparting a shiny and glamorous appearance. In addition, the liquid nature of the silicone resin enables excellent film forming properties to allow efficacy upon delivering to the hair. This resin is easily emulsified in both oil-based and water-based hair products, which further enhances the potential of this material. Other alteration of the functional groups may also inject various functionalities. Ionic interactions may be enhanced, hence increasing hydrophilic properties, capabilities to function as a thickener and improved percentage of deposition on the hair.[7]

Specialised hair products containing silicones may also offer thermal protection from plausible damage caused by heated styling tools such as curling irons and hair dryers. These tools tend to drive off moisture from the hair and lead to split hair. Silicones are able to remedy this as the material conducts heat gradually. The silicon film on the hair shaft moderates moisture loss, hence reducing hair damage. By reducing moisture penetration in and out of the hair cortex, the optimal moisture level can be maintained, which is essential for good hair strength.

Based on tests conducted on hair products, silicones are also found to help protect hair that has been treated with colorants by reducing discoloration. Normally, the color remains on the upper layers of the hair cuticle and may be easily washed off. Silicone has good spreading properties, which helps the colorants to effectively penetrate the hair cuticle, contributing to better colorant retention. Furthermore, silicone forms a water-insoluble film along the hair cuticle, which prevents discoloration.[8]

8.2.2 Silicone and Skin Care

Silicones in skin care have also prevailed significantly in the personal care market. The change in demographics around the world has driven consumers to desire additional products for anti-aging and beauty purposes to look young and radiant. The skin care industry uses this opportunity by developing skin care products that prevent premature aging, reverse certain aspects of aged skin and enhance the appearance of the skin.

Akin to hair products, dimethicones are also widely used for skin care products due to their emollient characteristics and their capacity to enhance skin feel for many skin care formulations.[4] Likewise, cyclomethicones are also used in many skin care products due to their low viscosity and volatile properties. This makes them suitable as delivery vehicles for many active ingredients in skin care. Particularly for dimethicones, their versatility to adhere various functional groups to the silicone backbone enabled many derivatives that produce differentiated properties. For instance, by grafting hydrophilic polyethylene oxide chains to the siloxane backbone, a non-ionic silicone surfactant may be formulated. This may be used in foam stabilisers, emulsifiers or wetting agents. This once again shows us the potential silicone products provide.

8.2.2.1 Sunscreen Products

Research has shown that one of the biggest contributions to skin aging is exposure to harmful UV radiation.[9] Studies have shown that this radiation results in cumulative skin damage that ultimately contributes to uneven pigmentation and loss of elasticity of the skin. This renders the importance of sunscreen products, which protect the skin from over exposure to UV radiation. Apart from protection from the UV radiation, sunscreen products are required to have a uniform film to maximise the protection. Silicones are known to help enhance the sun protection factor (SPF) of the product. This means that less sunscreen would be needed to obtain the same level of protection, enabling more formulation flexibility, reduced cost and a wider range of products offering UV protection.

Sunscreen products should also be able to withstand a certain degree of friction so that the film does not get removed easily by brushing on other surfaces. In addition, sunscreen products intended for sports or beach purposes should be resistant to water. Silicone addition increases the wash-off resistance, translating to longer-lasting protection from water and sweat.[10] Many consumers today are also particular about the product feel, which becomes a concern for many formulators. Fortunately, many silicone-containing sunscreen products today are able to provide both protection and a comfortable feel on the skin.

New advancement in silicon technology focuses on silicon-in-water emulsions that introduce novel sunscreens such as methylene bis-benzotriazolyl tetramethylbutyl phenol in combination with various vitamins and silicone elastomers. In addition, silicone elastomers make it plausible to formulate a transparent anhydrous elastomer gel incorporated with sunscreen formulas.[11] This new technology could create a breakthrough in sunscreen products, going beyond mere protection and providing enhanced nutrients to the skin.

Moreover, silicone fluids are often used in sunscreen products as they have the capability to decrease the oily and sticky feel on the skin. This is especially important for organic sunscreens due to their oily nature. To increase the competence of sunscreen products, various formations include phenyl silicone groups as they have better solubility in sunscreen oil and help to produce a more uniform film on the skin. On the other hand, silicone waxes that consist of long-chain alkyl groups grafted on the siloxane backbone help to increase the protection from the sunscreen, hence increasing the SPF value. This type of silicone wax exists in solid forms and it increases the viscosity of the sunscreen such that it remains on the skin longer.

8.2.2.2 Facial and Body Cleanser

Many existing organo-functional siloxane polymers are also widely used in facial and body cleansing uses. Volatile (poly)dimethylcyclosiloxane fluids are very commonly used in non-rinsable makeup removers and cleansers. Being good solvents for organic-based oils, they help to provide a smooth, dry and non-greasy feeling on human skin. The polymer is mild and gentle on the skin, which helps to reduce unwanted irritation, hence making it a suitable choice for its purpose.

Studies on water-soluble polydimethylsiloxane polyether brought about various new formulations for facial washes with better stable foams and skin-feel properties. In addition, the technology provides prospects of reduced eye and skin irritation. The spreadability of skin products could be altered through reducing the surface tension of water through the use of the newly developed polymer.[12] Furthermore, polyether functional siloxanes with high HLB (hydrophilic lipophilic balance) have potential benefits in surfactant-based systems as they leave behind a perceivable softness to the skin.

8.2.2.3 Daytime Hand and Body Skin Cream

To demonstrate the extensive use of silicones in skin care products, here are some commonly used formulas. Formula 1 in Figure 8.3 exhibits the use of cyclomethicone and dimethiconol. In this case, it's a blend of 13 percent high molecular weight PDMS polymer gum in cyclomethicone. This aims to improve the non-greasy skin feel upon application to the skin. Once the formulation evaporates or dries, the residual dimethiconol helps to produce a resultant smooth skin feel. The percentage of dimethiconol could be altered to suit the specific needs of the user. Dimethiconol in relatively small amounts (0.45%) can provide a drier and smoother feel upon application of the cream on the skin.[12]

8.2.2.4 Nighttime Skin Lotion

Formula 2 in Figure 8.4 illustrates how silicone emulsifiers (dimethicone copolyol) may be used to formulate basic water-in-cyclomethicone emulsion cream that produces an elegant after-feel. In this formulation, amodimethicone (Figure 8.5) is used as the dimethicone copolyol.

Hand and body face cream	
Water	84.7 %
Carbomer 934	0.1 %
Glycerin	3.0 %
Triethanolamine	0.9 %
Cetyl Alcohol	1.0 %
Stearic Acid	0.8 %
Glyceryl Stearate	1.5 %
Cyclomethicone (and) dimethiconol	3.0 %
Isopropyl myristate	1.5 %
Diisopropyl adipate	1.5 %
Mineral oil	2.0 %
Preservative/fragrance	q.s %

Figure 8.3 Formulation of a certain hand/body cream.[12]

Formula 2 Nighttime Skin Lotion (44)	
Cyclomethicone (and) Dimethicone Copolyol	10.0%
Cyclomethicone (and) Dimethiconol	10.0
Cyclomethicone	10.0
PPG-3 Myristyl Ether	0.5
Water	62.5
Glycerine	5.0
Sodium Chloride	2.0

Figure 8.4 Formulation of nighttime skin lotion.[12]

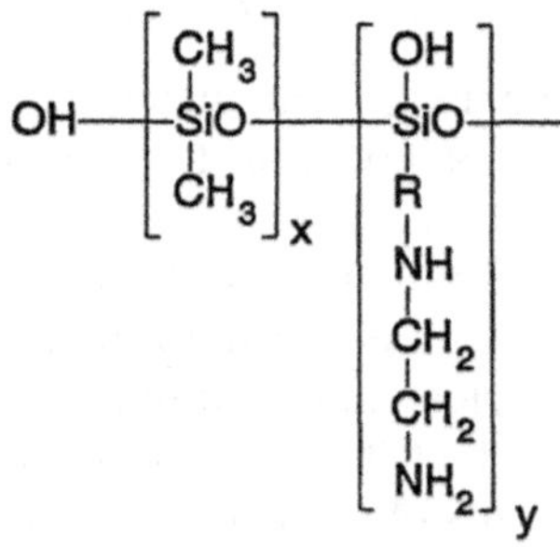

Figure 8.5 Chemical structure of amodimethicone.[12]

In many other formulations, cyclomethicone is often utilised in cleansers for makeup removal purposes. Furthermore, low molecular weight silicones are widely used in facial cleansers as they possess low surface tension, good wetting characteristics and the capacity to eliminate dirt or colour cosmetic residues, whilst providing a dry and non-greasy skin feel. In hygiene applications, the amount of foam produced is also often an important aspect for consideration. Most consumers favour products that produce a generous foam.[13] Attributing to their amphiphilic nature, silicone polyethers can control the water–air interface of the foam structure, contributing to a rise in stabilisation and volume of the foam generated by the cleansing surfactants of the formulation. Furthermore, some of these silicone polymers have been shown to significantly decrease the eye irritancy that can be caused by such anionic surfactants.

8.2.3 Silicone and Cosmetics

Beyond protection and cleansing, silicones are also frequently used for aesthetic purposes such as lipsticks, nail polishes and other cosmetic products. The extensive range of such products has enabled silicone industries to benefit significantly in the market. Based on studies from Cosmetic Research International, a large percentile of novel products introduced in 2002 contained silicone in their formulation.[14] Amongst the many different silicones, bis-hydroxyethoxypropyl dimethicone, a silicone carbinol fluid, is found to have great versatility. It functions as a wetting agent in addition to fragrance retention and moisturisation.[15] Bis-hydroxyethoxypropyl dimethicone has low viscosity and is hydrolytically stable. Hence, it is often easily included in cosmetic formulations that may include organic esters, alcohols and resins. Sometimes known as silicone carbinol fluid or carbinol-functional silicone, this polymer exhibits a unique structure. The small organic component offers a noteworthy polar and nonpolar compatibility profile. The structure further imparts hydrophilicity and wetting properties. Silicone fluids often have a small surface tension, which enables them to be excellent wetting agents and exhibit good spreadability. For instance, PDMS has a lower liquid surface tension compared to the critical surface tension of wetting. This justifies why the polymer spreads over its own adsorbed films. The spreadability of silicone polymers also depends on the compatibility with other materials in the formulation.

On a specific surface, a coating may spread or withdraw, depending on the surface tension. When the surface tension of the substrate is lower than that of the coating, the coating withdraws on the substrate. This helps to reduce common surface with the substrate. Bis-hydroxyethoxypropyl dimethicone (Figure 8.6) can be utilised to decrease the differences of the interfacial tension between coating and substrate, hence reducing the flaws in the surface constitution of the film. Similarly, the material may be used as a wetting agent for preparation of pigment dispersions.[14]

As much as the fluid provides both polar and silicone functionalities, its characteristics differ from many existing polar silicone polymeric materials such as silanol fluids and silicone polyethers. This material further shows excellent properties as a binder for powdered cosmetics, without the need to use various formulations typically used to bind pigments. In addition, colour cosmetics today are required to provide enhanced benefits to users. Due to its ethylene oxide moiety, the silicone carbinol fluid is hydrophilic, enabling it to bind water from the surrounding environment. Based on a study conducted by Dow Corning Corporation, silicone carbinol fluid may increase the hydration level of the skin approximately 10 to 20 percent across a time period of five hours. These hydrating capabilities may enable formulators to produce colour cosmetics with prolonged moisturisation properties (Figure 8.7).[15]

PDMS polymers are also important in the cosmetic sector as they have been known to be noncomedogenic and nonacnegenic, indicating they are not expected to encourage undesired skin pore clogging or acne. This is essential in many cosmetic products as consumers would not purchase a product that is detrimental to their skin. Lipstick is arguably one of the most

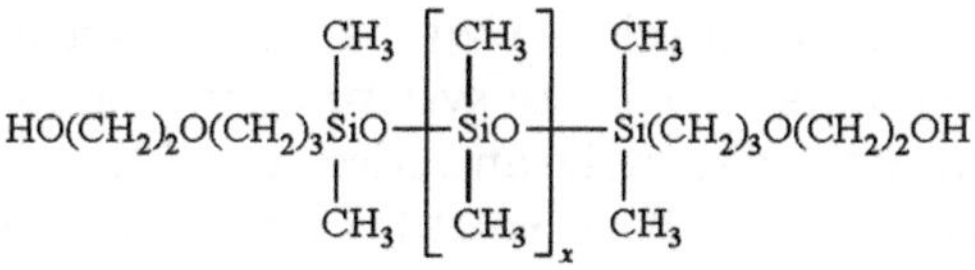

Figure 8.6 Bis-hydroxyethoxypropyl dimethicone.[15]

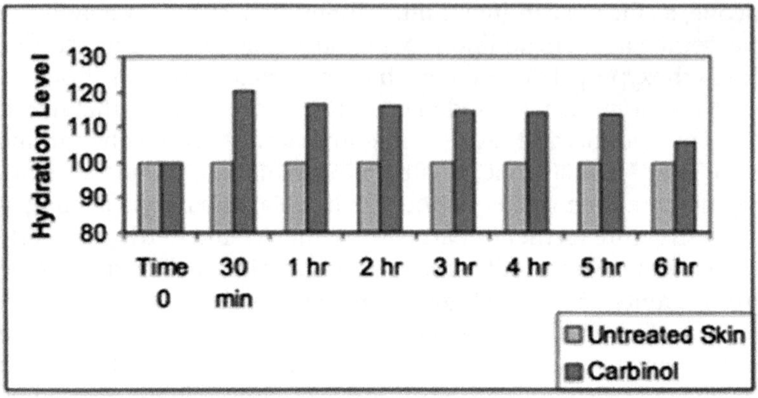

Figure 8.7 Hydration level of untreated skin *vs.* carbinol-treated skin.[15]

enduring cosmetic products around. Women and men around the world have coloured their lips at various times across decades. Lipsticks today come in countless numbers of colours and properties. Many of these incorporate silicone to reap its benefits. For example, alkylmethylsiloxanes are used because of their capability to produce a waxy uniformity, a pleasant feel and ability to enhance compatibility with organic ingredients commonly used in such formulations.[4] What most consumers seek in a lipstick would be how lasting it is coupled with the shine it produces and the colour. High melting silicone waxes may be produced to create a firmer stick and improve kiss-proof or transfer resistance.[16] Additionally, other types of silicone, such as phenyl silicones, possess a high refractive index, which helps to create a glossy look on the lips. On the contrary, a matte effect may also be created through silicone elastomer dispersions. This is due to their specific effect on light scattering. Various companies, such as Wacker, create volatile silicone fluids that immobilise coloured pigments, hence deterring the colour from seeping into the users' lip crease. In essence, a pigment is a finely divided crystalline solid that helps to impart colour to a particular substrate.[17]

8.3 Silicon As Delivery Systems

Silicone technologies have evolved to possess new and innovative delivery systems for active ingredients associated with personal care. These novel technologies have brought about limitless high-performance and highly functional products through the combination of specific silicones and desired active ingredients. A delivery system is often known as any type of material that has the ability to make an active available for the desired site of action. This could range from basic volatile carriers for actives to more complex technologies such as encapsulation. Encapsulation may provide properties such as controlled or slow release of the active.[9] Silicone delivery systems come in many modalities. One basic example would be using a volatile silicone, such as low molecular weight cyclomethicone, as a vehicle to deliver antiperspirant actives. Another type of system highlights how silicone may be used to impart a synergistic outcome. This is similar to the conditioning effect mentioned above, achieved through the use of dimethicone copolyol and organic quaternary materials to enhance the product properties.

Another typical active delivery technique may be known as encapsulation with no release. This method helps to reduce negative effects from the active ingredient, such as skin irritation or penetration, and at the same time sustain the properties required for its purpose. An example of this technique includes pigment coating for colour cosmetics. The most widely used delivery system used in the personal care industry would be encapsulation. This system aims to protect actives that are sensitive to environmental factors, to transport irritating ingredients and to prevent actives from unwanted interactions with other ingredients in the formulation. The release of the actives may be triggered during application through the high shear rate resulting from rubbing the material on the user's skin. Sequential release and controlled release methods are the most complex amongst the different delivery systems. This method may be used to combine chemically incompatible actives, protect actives encapsulated, and exhibit long-lasting, functional properties on the skin. The entrapment method enables complete release of an ingredient at a specific time whereas controlled release allows the ingredients to be delivered slowly over time. On the other hand, a sequential release method may isolate various actives until specified times for a longer lasting and effective outcome. For instance, multiple-phase emulsions may be used to deliver an array of active ingredients. Water-in-oil emulsions based on silicones such as polyether-modified elastomers or alkyl dimethicone copolyols may be used to deliver vitamin A and vitamin E. Other actives such as vitamin C could be added into the external aqueous phase that surrounds the emulsion containing the other actives.[18]

8.3.1 Silicone Vesicles

Liposomes are commonly known to form vesicles but silicone technologies has become a possible alternative. Silicone polyethers such as dimethicone copolyol are capable of forming vesicles that have surfactant-based structures. In addition, they may also display self-organisation abilities. For instance, surfactants may form various types of aggregates in water through the rearrangement of the group structure to increase the stability and obtain the lowest energy configuration possible. The structure of silicone vesicles may allow many essential active ingredients to be incorporated and delivered to the targeted site. For instance, by using silicone vesicles, the hydrophobic and hydrophilic actives may be separated from each other. Due to a bilayer structure, the two phases may be protected by each other. This separation may be achieved when the hydrophobic actives are dispersed in the bilayer whereas the hydrophilic actives can be in the solution, as a water phase. Both water-soluble actives and non-water soluble actives may be incorporated into the silicone vesicles. Some examples of these actives may be found in Table 8.1.

Table 8.1 Examples of water-soluble and non-water soluble actives.[18]

	Water-soluble actives		Non-water soluble actives
Conditioning agents	• Vitamin C	• Hydrolyzed collagen	• Vitamin A
	• Vitamin H	• Gelatin	• Vitamin E acetate
Preservatives	• Salicylic acid		• 2-Mercaptopyridinel-oxide
Sunscreen agents	• 4-Aminobenzoic acid (PABA) • 2-Phenylimidazol-5-sulfonic acid		• Homomethyl salicylate • 4-Methoxy cinnamic acid isoamyl ester
Drugs	• Nicotine and aspirin		• Nitroglycerin
Humectants	• Glycerin • Propylene glycol		• Lanolin alcohol • Cetearyl octnoate

Advantages of using silicone vesicles compared to traditional liposomes may include an increased stability as they do not contain hydrocarbon chains and silicone copolymers rearrange into vesicles spontaneously. Silicone vesicles are also very stable at relatively high temperature. For cosmetic purposes, the material has a good tolerance for many active ingredients in the respective formulations. Silicone vesicles having the ability to encapsulate actives provide many other benefits:

- controlled release of desired actives,
- reduced skin irritancy that may be caused by actives such as preservatives or sunscreens,
- enhanced stability of actives such as vitamins or enzymes,
- helps to incorporate various incompatible actives together.

The material's versatility has therefore brought about much attention, not only in personal care applications but numerous other applications in science today.[19]

8.4 Silicones and Their Properties

We have seen many different silicone derivatives and how they are altered to suit the specific needs of the personal care products. What makes some silicone products more widely used than others relies largely on the properties they possess.

8.4.1 Polydimethylsiloxane

As we have seen, PDMS (dimethicone) is one of the most commonly used silicones for personal care applications. It is imperative to investigate what contributes to the unique properties possessed by this polymer. This polymer may be broken down into two major components: the polar inorganic backbone and the nonpolar organic substituents. The polar inorganic backbone consists of the long and strong Si–O bond with a relatively open Si–O–Si angle. First and foremost, comparing the fundamental elements in silicones and hydrocarbon, specifically silicon and carbon. Silicon and carbon belong to the same group in the periodic table but they are quite different in many aspects:

- The size of silicon atom is larger than that of carbon atom.
- Silicon (1.8) is less electronegative than carbon (2.5). This means that the element is able to give up more of its electrons to other elements, which helps create stronger and more energetic bonds.
- Silicon bonds are longer than carbon bonds.
- Silicon has more flexible bonds with wider bond angles.
- Silicon forms only single bonds, which are more stable and resistant to breakage, while carbon may form multiple bonds.[20]

To further explain the potential of silicone in personal care applications, we may examine respective factors of the silicone structure. Firstly, Si–O has a bond energy of 452 kJ mol^{-1} while the C–O bond has a bond energy of 358 kJ mol^{-1}.[21] Furthermore, the Si–O bond is relatively longer and flatter than the C–O bond. Consequently, the structure exhibits a low barrier to rotation with low rotation energy required.[20] This combination of attributes explains why silicone material, including PDMS, possesses excellent flexibility and internal mobility as well as high free volume. This allows the respective functional group to obtain the optimal alignment configuration. The structure decreases competition amongst functional groups and reduces functionality requirements. Therefore, PDMS polymers may be more cost effective as in rigid organic polymers a higher amount of expensive functional groups are needed instead. This makes PDMS very suitable for personal care purposes.[22]

PDMS is highly chemically stable due to its siloxane backbone with a high bond energy of approximately 445 kJ mol^{-1} alongside the inert methyl functional groups attached to it. The polymer is resistant to temperature extremes, oxidation, moisture and many chemicals. Moreover, having a low surface tension of 20.4 mN m^{-1}, the polymer wets most surfaces easily. The attached methyl groups are aligned in the most optimal position to generate water-repellent films as well as excellent release properties. PDMS film forming abilities further boost its competiveness in the market as it outperforms hydrocarbons in creating extremely thin self-levelling films.

PDMS is also very shear stable. Due to low internal friction, PDMS is able to spread and flow much more efficiently than same-viscosity hydrocarbon fluids in many respective mechanical environments. Furthermore, PDMS is able to perform across a wider temperature range than many hydrocarbon fluids, from below 40 °C to above 150 °C. This is owing to its extremely low glass transition temperature and less temperature-dependent viscosity.[22]

8.4.2 Cyclomethicone

Another commonly used silicone for personal care products as mentioned is cyclomethicone (Figure 8.8). Being a volatile methyl siloxane, cyclomethicone belongs to a class of liquid silicone that has high volatility and low viscosity. Although the fundamental elements in the polymer are the same, cyclomethicone differs from PDMS whereby cyclomethicones are cyclic whereas PDMS is a linear siloxane polymer. PDMSs have a long backbone surrounded by various methyl functional groups, which creates a liquid that is clear and non-volatile. Cyclomethicones, on the other hand, have short backbones that have generally closed ring structures, giving them similar properties to PDMS but much more volatile. This renders cyclomethicones more relevant to some personal care products when complete evaporation of the siloxane carrier is required.

It is noteworthy that, although cyclomethicone is both cyclic and volatile, not all cyclic silicone structures are volatile. A cyclic silicone may be denoted by the structure shown in Figure 8.9.

If $n = 4$, the product is known as D4 (octamethylcyclotetrasiloxane); if $n = 5$, the product is known as decamethylcyclopentasiloxane. These polymers are volatile. However, higher values of n will deem the polymer nonvolatile.

As seen in Table 8.2, volatile silicones often translate to polymer materials that possess a low heat of vaporisation and have the ability to select a desired vapour pressure. This group of silicones can therefore evaporate from the skin efficiently, leading to a dry feel. D4 polymer will evaporate in approximately 30 minutes at 25 °C and within 10 minutes at 37 °C.[24] They may be

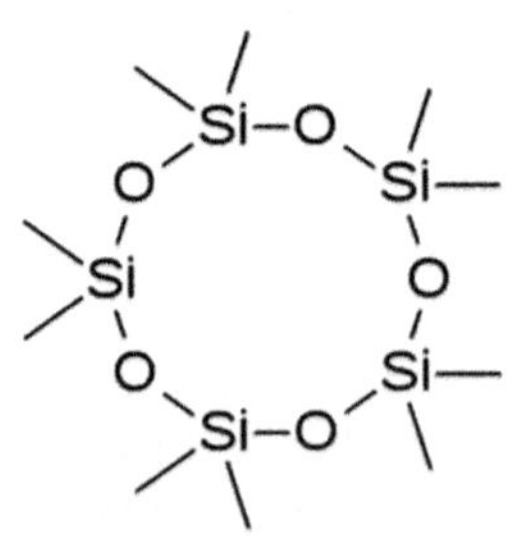

Figure 8.8 Chemical structure of cyclomethicone (D5).[19]

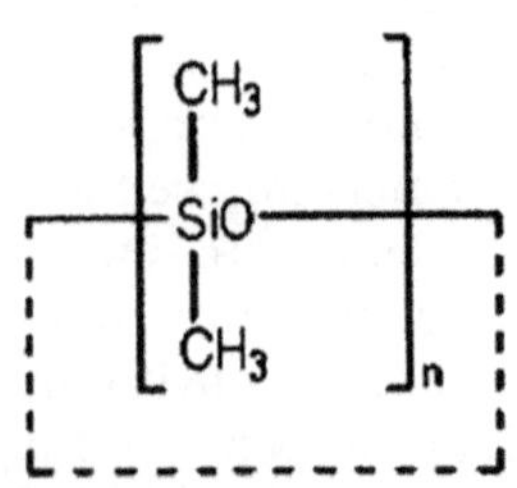

Figure 8.9 Chemical structure of cyclic silicone.[23]

Table 8.2 Heat of vaporisation for some volatile fluids used in cosmetics.[4]

Fluid	Heat of vaporisation (kJ kg^{-1})
PDMS, cyclic (DP = 4)	172
PDMS, cyclic (DP = 5)	157
Hexamethyldisiloxane	192
Ethanol	840
Water	2257

Table 8.3 Water vapour permeability for silicones.[4]

Fluid	Water vapour permeability (g m^{-2} h^{-1})
PDMS, cyclic (DP = 5)	155.7
PDMS, linear	107.4
Silicone gum	148.6
Silicone resin	110.5
Mineral oil	98.0
Alkylmethylsiloxane (C30+)	1.4
Petrolatum	1.3

either linear or cyclic. For example, hexamethyldisiloxane (MM) is a low viscosity silicone and volatile fluid even though it is not cyclic.[25]

Hence, we can see that cyclomethicone has established its foothold in many current and upcoming personal care products. Owing to its low heat of vaporisation, the polymer generally evaporates at a faster rate compared to conventional PDMS (dimethicones). For these reasons, it explains why cyclomethicone is more widely used in antiperspirant and deodorant formulations.

Cyclic silicones and linear PDMS (dimethicones) are also known to have excellent permeability. This translates to better breathability of the product which is often essential for skin care purposes. From Table 8.3, we may compare the respective water vapour permeability of the different families of silicones. Permeability is related to both solubility and diffusion coefficient. Silicones are permeable, as they possess a relatively high solubility for many different gases. Furthermore, silicones exhibit relatively higher gas diffusion rates in comparison with other commonly used polymers. Varying the substituted methyl groups may alter the amount of "breathability". For instance, by introducing longer alkyl groups, the permeability may be decreased. This is required when there is a need to retain skin moisture and plasticisation.[4]

8.4.3 Silicone Elastomers

In addition to conventional silicone fluids, there are also other classes of silicone such as silicone elastomers. Silicone elastomers are commonly known as silicone crosspolymers due to their high density of cross-linking. Many silicone elastomers are based on cross-linked dimethicone and presented as a swollen gel. Silicone fluids do not exhibit cross-linking, which contributes to their high flexibility and degree of freedom. As a result, silicone fluids remain in the liquid form even if the polymer possesses high molecular weight. On the other hand, silicone elastomers possess less freedom for rotation due to their highly cross-linked density, which in turn makes the polymer more stiff. Therefore, silicone elastomers may be used as thickening agents in personal care products. Being compatible with various lipophilic active ingredients, they may also function as stabilising agents. Hence, silicone elastomers are able to provide a different set of characteristics not seen in silicone fluids.

8.4.4 Silicone Resin

Another major class of silicones will be the silicone resins. This family of silicones is made up of a highly cross-linked, three-dimensional silicate structure. Unlike elastomers, silicone resins do not usually contain silica fillers. The structure enables them to function as a plasticiser in many personal care products. Silicone resins with high methyl content generally offer great flexibility and water resistance. Having excellent weathering properties, the polymers are often used as protective films with long lasting effects. In some hair care products, silicone resins are used to help generate volume and body, and simultaneously impart humidity resistance for the hair.[26] Having various distinct features, it is important to note that the polymers are not mutually exclusive. More often than not, formulations in personal care products incorporate silicones from different classes. These blends of silicones have helped formulators to generate products that inculcate properties from the different polymers, hence producing numerous valuable specialised products.

8.5 Issues with Silicone

Many personal care industries have established the benefits of silicones. It is intriguing why there are an increasing number of products labeled as "silicone-free." Many critics around the world attempt to stop the use of silicone in personal care products. Hence, it is imperative to investigate what the issues are that silicone products seem to raise.

8.5.1 Safety Issues

With an increasing abundance of silicone products in the market, consumers and health activists are becoming more concerned about the safety of the material. Ingredient activists such as Campaign for Safe Cosmetics attempt to put silicone in a bad light. It was claimed that cosmetic grade silicone would irritate the skin, cause allergic reactions and clog the pores. Research began to debunk the myth by clarifying that synthesised cosmetic grade silicones are not allergenic to the skin. In addition, the silicones used in personal care applications are too large to enter or clog a pore. Silicone fluids, being known to have a relatively open structure, make the material "breathable." This would mean sufficient air would be able to penetrate through, which is essential for cosmetic purposes. The American Academy of Dermatology found cosmetic grade silicone to be a beneficial ingredient in makeup applications such as foundation. Cystic acne and rosacea patients may reap the benefits of silicone-based makeup as it may reduce skin redness, stinging or irritation.[27] In Europe, cyclomethicone (D4) has been found to be an endocrine disruptor. Animal studies suggest that D4 exhibits toxicity profiles that may have indirect or direct impact on human health. The study exposes rats to lifetime inhalation exposure concentrations of D4 in air. The several hundred parts-per-million concentration corresponds to an exposure of more than 1500 mg kg^{-1} day^{-1} for humans. This could be unrealistic for human exposure as we do not come into that much contact with the polymer.[28] Furthermore, it has been said that screening assessment takes into consideration human exposure from uses of various silicones in personal care products and cosmetics, but no scientific evidence has showed they indeed pose a risk to human health at the current amount of exposure.[29] A study conducted found that D5 is in the largest quantity in many personal product applications and ranges from 0.01% w/w in hand creams to more than 35% w/w in various deodorants. Lower concentrations of D4 were found to be present in many applications. Sparse presence of D4 found in the study shows that many industries are phasing out D4 and substituting it with the less toxic D5 and D6. Hence, as far as silicones may produce a plausible risk to human health at very high concentrations, the personal care industries had been undertaking a role to ensure that exposure to silicone does not pose any detrimental effects to consumers.[28]

8.5.2 Environmental Concerns

From the respective research, silicones are deemed to be safe for use on human skin. However, critics of the material go on to investigate the possibility that silicones cause detrimental effects on the environment through bioaccumulation. Many environmental activists advocate consumers to avoid using silicone-containing products. In 2008, an Environment Canada Review found that certain cyclic silicones (D4 and D5) could harm the environment and have potential risks in aquatic organisms. Departments such as the Canadian Cosmetic, Toiletry and Fragrance Association (CCTFA) now strive to produce reasonable risk assessments to highlight potential environmental concerns and work out how they might be reduced.[29] Personal care formulators are also becoming more aware of the environmental concerns and have begun to work out more eco-friendly formulations.

8.6 Conclusion

We have seen the vast potential of silicone in personal care applications ranging from old cleansing soap and shampoos to new technologies that enable controlled release of specific actives. Personal care may be broken down into three main categories for hair products, skin products and cosmetic products. Nonetheless, silicones have a part to play in each category, tapping into the versatility of their polymer chain. With increasing demand for the enhanced benefits that may be derived from personal care products, we have seen more introductions of silicone polymers into our daily lives. As we have discussed, dimethicone, cyclomethicone and many other silicone derivatives each have their own unique characteristics to suit the needs and demands of various products. Cyclomethicone is known to be volatile while some other products are able to adsorb to the human skin or hair for a longer period of time. Silicones have revolutionised the application of many products. Many factors such as spreadability, colour, moisturising ability, sun protection and even repair of the hair or skin are possible today due to the introduction of this excellent polymer and its technologies. Silicone elastomers have also been created which are able to help provide anti-aging properties. Hence, by studying and looking into the structure and functional groups of various silicone polymers, we may investigate how the different polymers provide the different characteristics. Having more silicone in the market, it is also important to keep in mind the pros and cons that may arise from use of the product. With the ever-increasing usage of silicones, it is plausible that it might evoke environmental issues. Hence, it is the formulators' responsibility to ensure that silicone products are indefinitely safe for personal use as well as being environmentally friendly. Nevertheless, the silicone market continues to grow and deliver novel products that benefit the industry. In years to come, silicone could eventually be the key ingredient to our beauty and everyday care.

References

1. American Chemistry Council, *Silicone Uses*, http://sehsc.americanchemistry.com/Silicone-Uses.
2. Wikipedia, 2015.
3. Albright, *Silicone Rubber*, http://albright1.com/siliconerubber/.
4. J. L. Garaud, *Silicones in Personal Care Applications*, Dow Corning, Belgium, 2004.
5. A. Urrutia, *Silicone: The Basis of a Perfect Formulation for Hair Care*, Dow Corning Technical Notes, 2001.
6. I. R. A. Geoffrey Hawkins, J. H. Xavier and L. C. Popescu, Cosmetic Compositions Containing Thiomers For Hair Color Retention, US Pat., US20110229430, 2009.
7. K. Schaefer, http://www.cosmeticsandtoiletries.com/formulating/function/moisturizer/7653302.html.
8. F. L. T. O. B. Johnson, T. Leaym, J. DeCaire, K. Quackenbush, *New Silicone Technologies for Ethnic Hair Care*, Dow Corning Technical Notes, 2012.
9. M. S. Starch, *Silicones' Benefits in Antiaging Skin Care*, 2008.
10. T. C. S. D. Michael Starch, I. Vervier, I. Van Reeth and M. C. T. Ramos, *Expanding Silicone Technologies for Sun Care*, Dow Corning Technical Notes, 2007.

11. I. V. Reeth, *Silicones Offer Versatility To Meet Global Personal Care Trends*, http://www.happi.com/contents/view_features/2011-02-17/silicones-offer-versatility-to-meet-global-pe/.
12. *Conditioning Agents for Hair and Skin*, ed. P. R. Randy Schueller, Marcel Dekker, America, 1999.
13. J. Blakely, *The Benefits of Silicones in Facial and Body Cleansing Products*, Dow Corning Europe, Form 22-1549-01, 1996.
14. P. s. C. Skincare, *Silicone*, http://www.paulaschoice.com/cosmetic-ingredient-dictionary/definition/silicone.
15. H. M. V. D. Adriana Urrutia, G. Brissette, R. H. Paul Pretzer, I. Van Reeth and V. Caprasse, *Silicone Carbinol Fluid*, Dow Corning, Form 27-1253-01, 2006.
16. N. R. a. N. W. O. H. Rachel Mallon, *Lipsticks: How They Have Changed and Where They Are Going*, http://www.cosmeticsandtoiletries.com/research/chemistry/premiumLipsticks-How-They-Have-Changed-and-Where-They-Are-Going–239944541.html.
17. Wacker, *Simply Beautiful: A Guide to Silicones for the Cosmetics Industry*, http://www.wacker.com/cms/media/publications/downloads/6022_EN.pdf.
18. C. S. Stephanie Postiaux, A.-M. Vincent, and J. Newton, *Delivery System Handbook for Personal Care and Cosmetic Products*, Dow Corning S.A., Seneffe, Belgium, 2005.
19. *Delivery System Handbook for Personal Care and Cosmetic Products: Technology, Applications and Formulations*, ed. M. Rosen, 23 September 2005.
20. Dow Corning, *Silicone vs. organic polymers*, http://www.dowcorning.com/content/discover/discoverchem/si-vs-organic.aspx.
21. J. Benson, *Wired Chemist*, http://www.wiredchemist.com/chemistry/data/bond_energies_lengths.html.
22. Dow Corning, *Physical and chemical properties of silicone (polydimethylsiloxane)*,http://www.dowcorning.com/content/discover/discoverchem/properties.aspx.
23. K. A. Bissah, Compositions for repelling fluid and uses thereof, US Pat., US20110152364, 2009.
24. E. D. Goddard and J. V. Gruber, in *Silicone in Cosmetics*, ed. J. V. Gruber and E. D. Goddard, Parcel Dekker, Inc., United States of America, 2nd edn, 1999, vol. 22, p. 688.
25. T. O'Lenick, *Cyclic Silicone vs. Volatile Silicone*, http://www.cosmeticsandtoiletries.com/research/chemistry/31318009.html.
26. R. Houben and B. Prinssen, Silicone crosspolymers: customizing sensory feel, *Personal Care*, 2011, http://www.personalcaremagazine.com/Story.aspx?Story=13149.
27. K. J. Bennett, *Cosmetic Silicone Safety: Fact vs. Fiction*, http://kjbennett.com/cosmetic-silicone-safety-fact-fiction/.
28. T. Dudzina, N. von Goetz, C. Bogdal, J. W. H. Biesterbos and K. Hungerbühler, *Environ. Int.*, 2014, **62**, 86–94.
29. A. Chandler, *Reader's Digest*, 2012.

CHAPTER 9

Towards Cyclodextrin-Based Supramolecular Materials

ANIS ABDUL KARIM[a] AND XIAN JUN LOH*[a,b]

[a]Institute of Materials Research and Engineering, A*STAR (Agency for Science, Technology and Research), 2 Fusionopolis Way, Innovis, #08-03, Singapore 138634, Singapore; [b]Department of Materials Science and Engineering, National University of Singapore, Singapore 117574, Singapore
*E-mail: lohxj@imre.a-star.edu.sg

9.1 Introduction

Inclusion complexation between cyclodextrins (CDs) and various guests has been extensively investigated in supramolecular chemistry. Besides CDs, there are several important macrocyclic host families, such as crown ethers[2,3] and cucurbiturils.[4,5] Until now, the contribution of these other families to macromolecular self-assembly has been small compared to CDs. This chapter will focus on CDs as hosts for interaction with guest monomers to form hydrogels. Hydrogels, a class of water-swollen crosslinked polymeric materials with characteristic three-dimensional network structures, are of increasing interest because of their unique properties and potential applications in drug delivery, tissue engineering, wound dressing, and cell immobilization.[6–12] Stimuli-responsive polymeric hydrogels have potential applications in many fields as "intelligent" materials because of their reversible responsiveness to many external stimuli, such as pH, temperature, light, or magnetic fields. Molecular assemblies organized by intermolecular forces are

RSC Polymer Chemistry Series No. 20
Polymers for Personal Care Products and Cosmetics
Edited by Xian Jun Loh

Published by the Royal Society of Chemistry, www.rsc.org

essential for the construction of a wide variety of these dynamic polymeric networks in aqueous systems *via* a host–guest system. Progress in supramolecular chemistry has paved the way for development of hydrogels formed based on the polymer–CD inclusion complexes (ICs).

CDs are cyclic oligosaccharides made of glucopyranose units linked by α-1,4 glucosidic bonds. The most common CDs are α-cyclodextrin (α-CD), β-cyclodextrin (β-CD), and γ-cyclodextrin (γ-CD), which consist of six, seven, and eight glucopyranose units, respectively. They present a truncated cone shape with a hydrophilic exterior composed of –OH groups, and a hydrophobic interior which can accommodate nonpolar compounds.[13–15] Hydrophobic and van der Waals interactions between the inner surface of the CD ring and the hydrophobic guests with a suitable molecular size for the CD cavity are responsible for inclusion complexation. CDs have been used as a cyclic component by molecular recognition with polymeric moieties in host–guest systems. Figure 9.1 shows the physical properties of α-, β-, and γ-CDs.[16]

It was found that α-, β-, γ-CDs were selectively threaded onto various linear polymers. Over the past few decades, inclusion complexation involving interactions of CDs and poly(ethylene glycol) (PEG)-based polymers has been widely employed in the construction of supramolecular hydrogels.[17–21] Supramolecular hydrogels formed by threading CD rings onto PEG-grafted copolymers have been reported,[12,22–24] including more complicated polymers such as PEG-grafted dextran,[25] PEG-grafted chitosan,[26] PEG brushes,[27] and star PEG polymers.[28] This chapter discusses CDs as cyclic host molecules and interaction with PEG-based guest molecules to construct supramolecular materials. As reported by Harada *et al.*, CD interaction with other monomers was made possible depending on proper molecular recognition.[16]

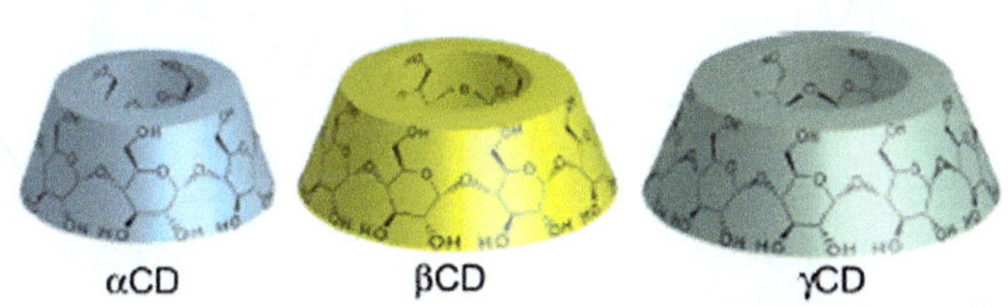

	αCD	βCD	γCD
No. of Glucose Units	6	7	8
Cavity Diameter (Å)	4.7	6.0	7.5
Height of Torus (Å)	7.9	7.9	7.9

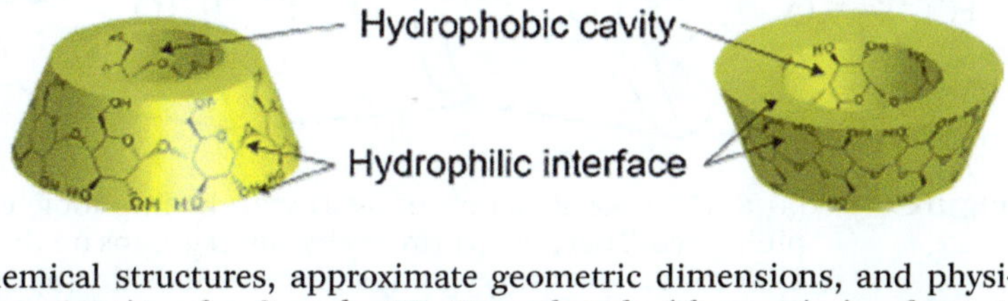

Figure 9.1 Chemical structures, approximate geometric dimensions, and physical properties of α, β, and γ-CD. Reproduced with permission from A. Harada, Y. Takashima and M. Nakahata, *Acc. Chem. Res.*, 2014, **47**, 2128–2140. Copyright (2014) American Chemical Society.[16]

Macroscopic molecular recognition can be categorized by three types of interactions: main chain (polyrotaxane), side chain, and sequential complexes.[16] Utilizing CD as host molecule, polymers such as polyethers, cationic polymers, polyamines, polyesters, π-conjugated polymers, polyolefins, polyamides, polyurethanes, and inorganic polymers could interact to form an IC. This chapter will attempt to discuss these studies.

Depending on the functional groups attached to the polymeric component, supramolecular formation can be altered based on the stimuli response.[1,29] Introducing polymer side chains or groups that respond selectively towards external stimuli could affect the hydrogel formation.

9.2 α-CD + PEG Gels

In the 1990s, Harada *et al.* successfully prepared ICs of α-CDs with PEG.[30–34] PEG was the first polymer found to form ICs with α-CD in the solid state (Figure 9.2). ICs with α-CD are only formed with tetra(ethylene glycol) and longer ethylene glycol polymers. The yield of the complex increases as the molecular weight of PEG increases, and complexes are obtained almost quantitatively with PEG when the degree of the polymerization is greater than 10.[30–35]

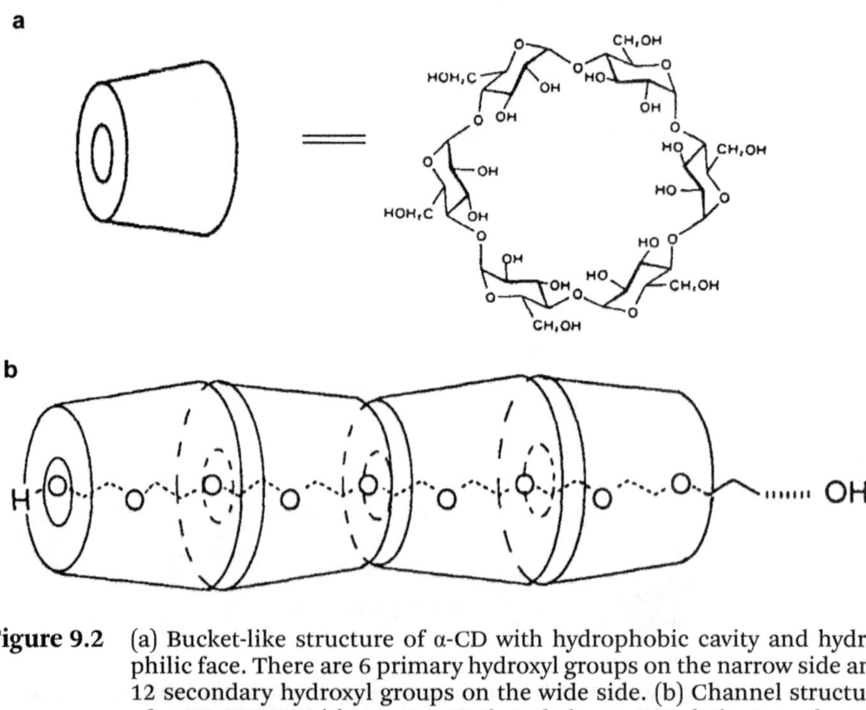

Figure 9.2 (a) Bucket-like structure of α-CD with hydrophobic cavity and hydrophilic face. There are 6 primary hydroxyl groups on the narrow side and 12 secondary hydroxyl groups on the wide side. (b) Channel structure of α-CD–PEG IC with many α-CD threaded on a PEG chain. In each α-CD cavity are bound two ethylene glycol units. Reproduced by permission from Macmillan Publishers Ltd: *Polym. J.* ref. 33. Copyright (1994).[33]

When α-CD saturated aqueous solution is mixed with PEG aqueous solution, the solution becomes turbid.[31] Inclusion compounds comprised of PEGs with high molecular weights (above 100 000) have yielded physical hydrogels where the crystalline parts are formed by the PEG complexes as the crosslinking parts. Heating the hydrogels results in homogeneous solutions, and this process is reversible.[33] IC formation between PEG and α-CD is affected by the mobility or shape of the terminals of the linear polymer. Harada *et al.* reported that steric hindrance for IC formation is changed in relation to the shape of the bulky-terminal groups introduced at the terminals of the polymer.[36] PEG modified with functional groups will affect the host–guest interaction and consequently the hydrogel characteristics, such as response to external stimuli.[19]

9.2.1 α-CD + Modified PEG Gels

Modifications of PEG with hydrophobic, biodegradable monomers impart new properties to existing material. Ye *et al.* demonstrated synthesizing a polyglycerol sebacate (PGS) with poly(ethylene glycol) methyl ether methacrylate (PEGMA) and forming a supramolecular hydrogel by interaction with α-CD. PGS-PEGMA/α-CD hydrogel exhibited a tunable, low upper critical solution temperature (UCST), low gelation concentration, rapid gelation, and rapid self-healing ability with relatively high modulus.[37] Kai *et al.* prepared a series of PEGMA-grafted lignin hyperbranched copolymers; due to their hyperbranched architecture, aqueous solutions were found to form supramolecular hydrogels with a very low critical gelation concentration of 1 wt% copolymers, in the presence of α-CD (Figure 9.3).[38] These green supramolecular hydrogels are potentially useful as a smart biomaterial for

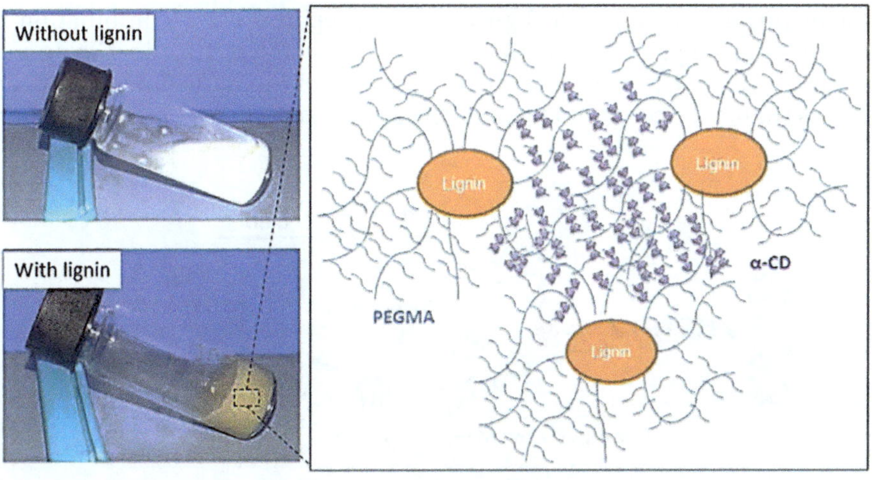

Figure 9.3 Illustration of lignin-PEGMA/α-CD hydrogel. Reproduced with permission from D. Kai, Z. W. Low, S. S. Liow, A. Abdul Karim, H. Ye, G. Jin, K. Li and X. J. Loh, *ACS Sustainable Chem. Eng.*, 2015, **3**, 2160–2169. Copyright (2015) American Chemical Society.[38]

biomedical application. Karim and Loh reported a supramolecular hydrogel made from a tri-component copolymer PLLA/DMAEMA/PEGMA and α-CD. Micelle association of the amphiphilic polymer with α-CD due to threading of PEGMA in the α-CD cavity formed ICs.[39]

As mentioned earlier in this chapter, an IC between PEG and α-CD is only formed with PEGs of high molecular weights (above 100 000). Guo *et al.* reported that adamantine (ADA) mono-end-functionalized low-MW PEG (ADA-PEG, M_n = 1.1 or 2K) promoted the gelation of the ICs with α-CD.[40] This strategy utilizes a hybrid inclusion complex (HIC) as a precursor to fabricate supramolecular hydrogels (Figure 9.4). HIC firstly formed between the silica particle coated with β-CDs and ADA-ended PEG. Then the free end of PEG threaded into the α-CDs, resulting in inter-HIC aggregation, and finally a hydrogel. The functionalized inorganic moieties could be regarded as "supercrosslinks" because each of the particles connects many polymer chains *via* supramolecular interactions.

Another study by Liao *et al.* demonstrated that end-modification of PEG with pyridinium (PEG-N^+) enabled the chains to form a brush-like conformation on clay surfaces.[41] This enabled PEG chains to thread into the cavities of α-CDs, leading to hybrid hydrogels. Tian *et al.* reported biodegradable poly(organophosphazenes) containing side chains of various oligo(ethylene glycol) methyl ethers (mPEGs) and glycine ethyl ester units. Novel supramolecular-structured hydrogel systems based on the IC between the mPEG grafted polyphosphazenes and α-CD were prepared in aqueous media (Figure 9.5). The gelation time depended on the length of the mPEG side chains, the molar ratio between mPEG repeat units and α-CD, and the concentration of the polymeric gel precursors. Polymers with longer PEG side chains had better stability and slower protein release profiles.[42]

Li *et al.* reported that an acid-labile and PEGylated polyphosphoester-doxorubicin prodrug (PBYP-g-PEG-g-DOX) was prepared *via* a combination of ring-opening polymerization (ROP) and Cu(I)-catalyzed azide–alkyne cycloaddition (CuAAC) "click" chemistry.[43] The resulting multifunctional prodrug was then used to interact with α-CD to fabricate a novel supramolecular

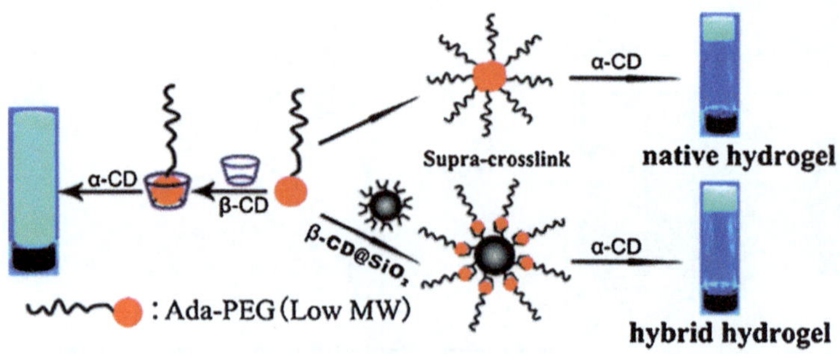

Figure 9.4 Schematic representation of the complex between ADA-PEG, α-CD, and β-CD-SiO_2. Reproduced with permission from M. Guo, M. Jiang, S. Pispas, W. Yu and C. Zhou, *Macromolecules*, 2008, **41**, 9744–9749. Copyright (2008) American Chemical Society.[40]

hydrogel based on inclusion complexation. Fast hydrogelation might result from a combination of the inclusion complexation between α-CD and amphiphilic block copolymer in aqueous solution. Zhao *et al.* demonstrated that supramolecular hydrogel formed from water-soluble amphiphilic poly(ε-caprolactone)-poly(ethylene glycol)-poly(ε-caprolactone) (PCL-PEG-PCL) block copolymers and inclusion complexation with α-CD (Figure 9.6).[12]

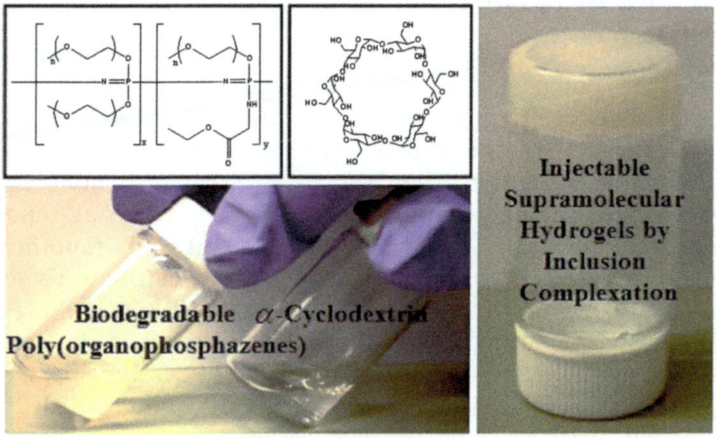

Figure 9.5 Illustration of mPEG grafted polyphosphazenes and α-CD hydrogels. Reproduced with permission from Z. Tian, C. Chen and H. R. Allcock, *Macromolecules*, 2013, **46**, 2715–2724. Copyright (2013) American Chemical Society.[42]

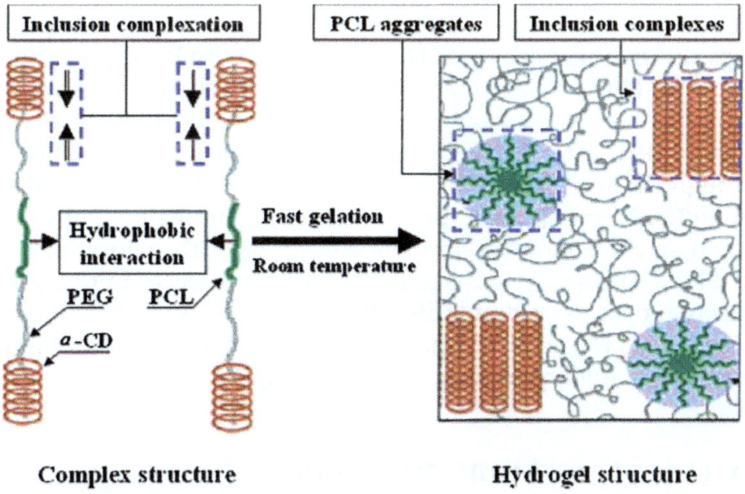

Figure 9.6 Graphical description of the gelation mechanism for the formation of the supramolecular hydrogel. Reproduced with permission from S.-P. Zhao, L.-M. Zhang and D. Ma, *J. Phys. Chem. B*, 2006, **110**, 12225–12229. Copyright (2006) American Chemical Society.[12]

When bulky-terminal groups are introduced to the terminals of the polymer, the steric hindrance for poly*pseudo*rotaxane (PPR) formation is changed in relation to the shape of the bulky-terminal groups. For instance, Harada *et al.* reported that bis(2-naphthylacetyl)-terminated PEG forms a PPR with α-CD, whereas bis(1-naphthylacetyl)-terminated PEG does not.[36]

9.3 γ-CD + PEG Gels

IC formation depends on the cavity size of CDs and the functional groups attached to the polymer. For example, negligible complexation occurs between γ-CD and PEG because the cavity size of γ-CD is too large for threading of PEG. However, recent studies have shown that modification of the structures aid in IC formation. Theoretically, γ-CD can form double-stranded ICs with PEG due to its larger internal diameter. Huang *et al.* reported the formation of double-stranded ICs between γ-CD and non-modified PEG.[44] The obtained γ-CD/PEG ICs showed a double-stranded structure whose yield, appearance, and forming rate were significantly affected by PEG molecular weight and temperature. Addition of a bulky group to the end of PEG capped the IC with γ-CD. Bis(3,5-dinitrobenzoyl)-PEG and bis(2,4-dinitrobenzoyl)-PEG form crystalline complexes with γ-CD in high yields, but α-CD does not form ICs with these PEG derivatives because the end groups are too large to thread into the cavities of α-CD. The authors have found that two ethylene glycol chains are included in the γ-CD cavity. Harada *et al.* studied PEG with fluorescent probes at the ends of the polymer chain and bis(1-naphthylacetyl)-PEG (PEG-1-N2) and bis(2-naphthylacetyl)-PEG (PEG-2-N2) complexes.[36] These complexes are composed of double chains of PEG, which penetrate the cavity of γ-CDs. Kong *et al.* reported a pentablock copolymer of 2-hydroxyethyl methacrylate (HEMA) end-capped with Br-poly(propylene glycol) (PPG)-PEG-PPG-Br.[45] Attaching PHEMA blocks altered the self-assembly process of the pentablock copolymer with γ-CDs in aqueous solution. Before attaching the PHEMA, the Br-PPG-PEG-PPG-Br was preferentially bent to pass through the inner cavity of γ-CDs to give rise to tight-fit double-chain stranded PPRs. After attaching the PHEMA, the resulting pentablock copolymer was single-chain stranded into the interior of γ-CDs to form more stable, loose-fit PPRs. The study indicated that γ-CDs can accommodate and slip over PHEMA blocks to randomly distribute along the entire pentablock copolymer chain. This results in unique, single-chain stranded PPRs showing no characteristic channel-type crystal structure (Figure 9.7).

9.4 CD + Other Monomer Gels

The major driving forces for forming CD ICs are hydrophobic and van der Waals interactions between the inner surface of the CD ring and the hydrophobic sites of the guests. Numerous studies on other monomer gels which incorporate CDs to form ICs were reported and will be discussed in this section.

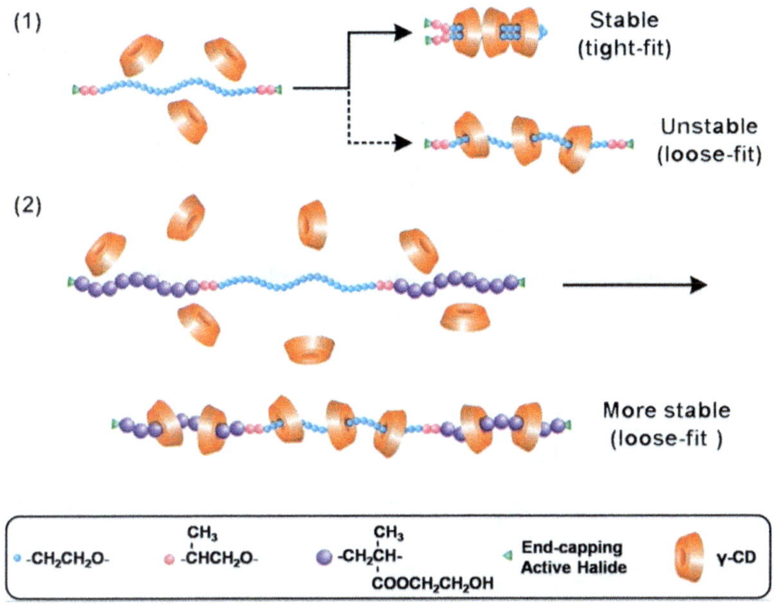

Figure 9.7 Schematic description of self-assembly of γ-CDs with Br-PPG-PEG-PPG-Br and PHEMA-PPG-PEG-PPG-PHEMA in aqueous solution. Reproduced from Kong *et al.*[45]

9.4.1 Polyethers

Harada's group discovered an efficient formation of PPRs with other water-soluble polyethers, such as poly(tetramethylene oxide) (PTHF),[34] poly(ethyl vinyl ether) (PEVE),[46] and poly(*n*-propyl vinyl ether) (PnPVE)[46] (Table 9.1).

α-CD does not form complexes with PPG regardless of the molecular weight.[34,35,47,48] Only γ-CD forms ICs with poly(methyl vinyl ether) (PMVE).[46,49] β-CD does not yield ICs with PEG but produces complexes with poly(propylene glycol) (PPG).[34,47,48] Although PEG does not form ICs with β-CD by mixing aqueous solutions, Udachin *et al.* reported a hydrogen-bonded head-to-head crystal packing structure of CD dimers with 3 monomeric units of PEG threaded through for every one β-CD.[50]

9.4.2 Polyamines

Poly(iminoundecamethylene), poly(iminotrimethyleneiminodecamethylene), polyethylenimine, and poly(ε-lysine) have been reported to form PPRs (Table 9.2). The formation of PPRs with polyethylenimines depends heavily on the pH of the aqueous medium. Although the maximum yield of PPR with α-CD and γ-CD has been observed at pH 11.0, complexation does not occur in the pH range below 8.0 due to protonation of the secondary amine groups in polyethylenimines. Poly(iminoundecamethylene) and

Table 9.1 Complex formation between CDs and polyethers. Reproduced with permission from A. Harada, A. Hashidzume, H. Yamaguchi and Y. Takashima, *Chem. Rev.*, 2009, **109**, 5974–6023. Copyright (2009) American Chemical Society[1]

			Yield (%)		
		$M_w/10^3$	α-CD	β-CD	γ-CD
PEG	$-(CH_2-CH_2O)_n-$	1.0	92	0	Trace
PTMO	$-((CH_2)_2-CH_2O)_n-$	1.0	94	47	Trace
PTHF	$-((CH_2)_3-CH_2O)_n-$	1.0	86	Trace	99
PPG	$-(CH_2-CH(CH_3)-O)_n-$	1.0	0	96	80
PMVE	$-(CH_2-CH(OCH_3))_n-$	1.0	0	0	80
PEVE	$-(CH_2-CH(OEt))_n-$	1.2	0	0	71
PnPVE	$-(CH_2-CH(OnPr))_n-$	1.2	0	0	4

Table 9.2 Complex formation between CDs and polyamines. Reproduced with permission from A. Harada, A. Hashidzume, H. Yamaguchi and Y. Takashima, *Chem. Rev.*, 2009, **109**, 5974–6023. Copyright (2009) American Chemical Society[1]

Polymer structures	CDs
$[NH-(CH_2)_2]_n$	α-CD, γ-CD
$[NH-(CH_2)_{11}]_n$	α-CD
$[NH-(CH_2)_{10}-NH-(CH_2)_3]_n$	α-CD
$[NH-(CH_2)_4-CH(NH_2)-CO]_n$	α-CD

poly(iminotrimethylene-iminodecamethylene) contain a long hydrophobic alkyl chain, which must be included in the α-CD cavity.

9.4.3 Cationic Polymers

Charged centers are necessary at both ends of the aliphatic chains to achieve dynamic stability.[51,52] CDs form inclusion compounds with ionic polymers[53–57] (Table 9.3).

Table 9.3 Complex formation between CDs and cationic polymers. Reproduced with permission from A. Harada, A. Hashidzume, H. Yamaguchi and Y. Takashima, *Chem. Rev.*, 2009, **109**, 5974–6023. Copyright (2009) American Chemical Society[1]

Polymer structures	CDs
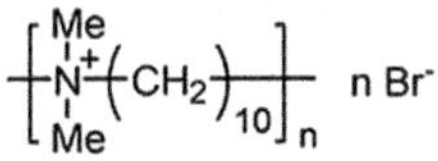	α-CD, γ-CD
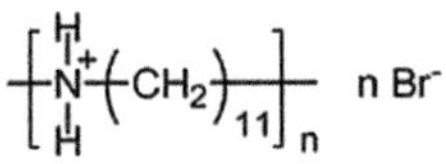	α-CD
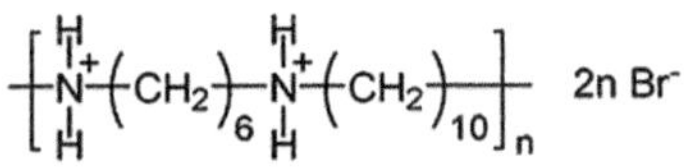	α-CD
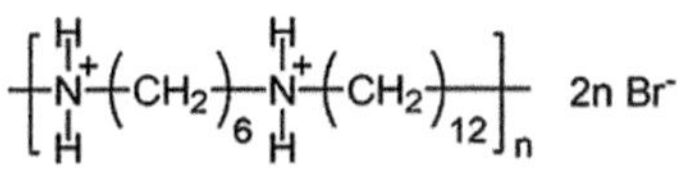	α-CD
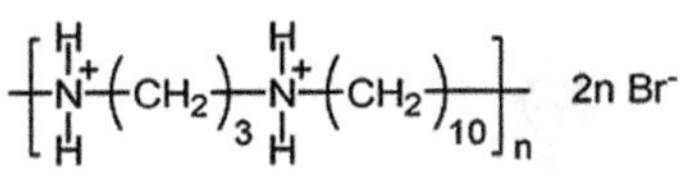	α-CD
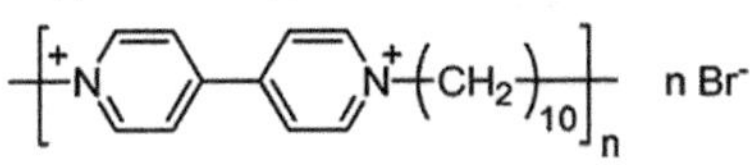	α-CD

Wenz *et al.* studied multiple cationic polymers forming ICs with α-CD and β-CD to give complexes soluble in water. One such study on kinetics of threading α-CD onto cationic and zwitterionic poly(bola-amphiphiles), which have long hydrophobic parts and hydrophilic groups (cationic groups), allows threading of CD rings under homogeneous conditions in aqueous solutions.[58] The resulting lower packing density of threaded CD rings compared to the packing density of channel inclusion compounds might be the reason for their better water solubility, because the polymer chain retains some conformational flexibility. Another study reported that mixing viologen polymer (VP) and the α,α-CD dimer creates a hydrogel, which is expected to realize supramolecular materials with a high tensile strength and self-healing abilities (Figure 9.8).[59]

9.4.4 Hydrophobic Polymers

CDs form ICs with hydrophobic polymers, such as polyethylene and polypropylene, with molecular weights less than 10 000.[31,35] γ-CD selectively forms ICs with poly(isobutylene).[60] The molecular weight selectivity of this polymer is reversed between β-CD and γ-CD; although β-CD forms complexes with low molecular weight compounds and the yields of the complexes decrease as the molecular weight increases, γ-CD gives complexes with this polymer and the yields increase with molecular weight. In addition, CDs form ICs with some polyesters, including poly(alkylene adipates)[61] and poly(ε-caprolactone).[62–64] In these cases, the polymer sample is heated with aqueous solutions of CD with vigorous stirring or agitation by sonication. The polymer sample is partly solubilized in aqueous solutions, and the complexes form as a solid state. In recent years, research on hydrophobic polymers such as polyesters and polyamides has been important due to their biodegradability and potential in pharmaceutical and biomedical applications. Formation and characterization of PPRs with polyesters are summarized in Table 9.4. The formation of PPRs is controlled by the chemical structure and stereochemistry, as demonstrated

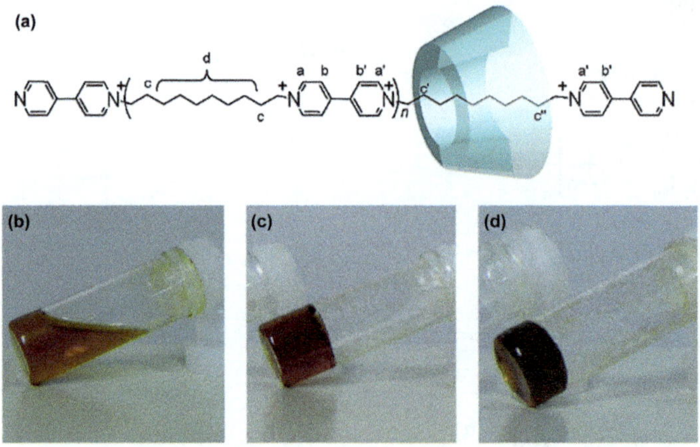

Figure 9.8 Supramolecular gels with self-healing and high tensile strength. Reproduced from Takashima *et al.*[59]

Table 9.4 Complex formation between CDs and polyesters. Reproduced with permission from A. Harada, A. Hashidzume, H. Yamaguchi and Y. Takashima, *Chem. Rev.*, 2009, **109**, 5974–6023. Copyright (2009) American Chemical Society[1]

Polymer structure		CDs
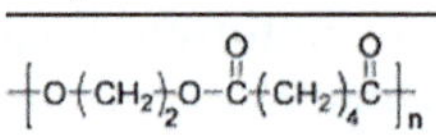	PEA	α-CD, γ-CD
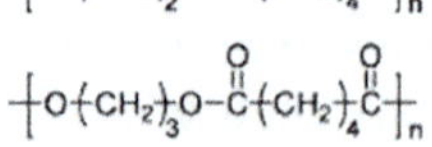	PTA	α-CD
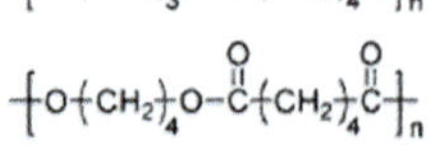	PBA	α-CD
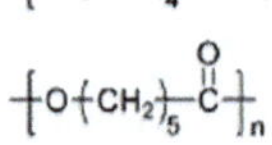	P(ε-CL)	α-CD, γ-CD
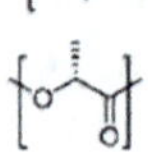	PLLA	α-CD
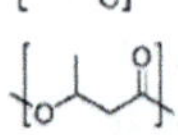	a-PHB	γ-CD
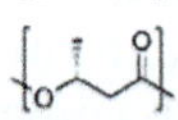	i-PHB	α-CD
P(ε-CL)-PEO-P(ε-CL)		α-CD, γ-CD
P(ε-CL)-PLLA		α-CD
P(ε-CL)-PPG-P(ε-CL)		α-CD, γ-CD

by Shuai *et al.* in stereoselective formation of ICs of poly(3-hydroxybutyrate)s with CDs.[65] Optically active poly((*R*)-3-hydroxybutyrate) (i-PHB) and atactic poly((*R*,*S*)-3-hydroxybutyrate) (a-PHB) selectively form crystalline ICs with different CDs. Another study demonstrated that when the PPR of polycaprolactones, P(ε-CL)-PPG-P(ε-CL) is prepared by α-CD, only P(ε-CL) blocks are incorporated in the α-CD cavity, whereas both P(ε-CL) and PPG blocks are included in the PPR of P(ε-CL)-PPG-P(ε-CL) and γ-CD.[66] These studies demonstrated that IC formation between polymers and CDs is governed not only by their chemical structures but also by their stereochemistries.

9.4.5 Block Copolymers

Site-selective complexation plays an important role in constructing supramolecular structures in biological systems and in chemical processes, and it should play a crucial role in constructing artificial supramolecular structures in polymeric systems. For example, Harada *et al.* reported that α-CD specifically binds an ethylene glycol, and β-CD binds the hydrophobic end group (an iso-octylphenyl group and a phenyl group).[67] Osman *et al.* reported a self-assembling hydrogel system based on the association of two polymers: polymerized β-CD and hydrophobically modified cholesterol-terminated PEG polymers (8-arm PEG-20k end-functionalized with cholesterol units and 8-arm PEG-20k end-functionalized with ADA units).[68] The inclusion complexation occurs between cholesterol moieties and β-CD cavities and the self-association of cholesterol moieties in aqueous solution. A modified system showed the suitability for the loading and release of bovine serum albumin (BSA) as a model bioactive drug. Star-shaped polymers could increase the number of binding sites to form strong hydrogels with respect to linear polymers. Hennink and coworkers reported a novel hydrogel system, based on cholesterol-ended 8-arm PEG with either free β-CD, or β-CD-ended 8-arm PEG.[69,70] The elastic modulus of the hydrogel, composed of the 8-arm β-CD-ended PEG as the host and 8-arm cholesterol-ended PEG as the guest, was much higher than that of the same host and linear cholesterol-ended PEG, at the same weight content. This result clearly showed the stronger gelation ability of the 8-arm structures.

9.4.6 Hybrid Inclusion Complex (HIC)

Introducing inorganic nanoparticles or nanosheets into supramolecular hydrogels could improve mechanical properties of these gels and render new functionalities. Earlier in this chapter, we discussed HICs of ADA-ended low molecular weight PEG. There are other studies reporting hybrid organic–inorganic hydrogels that have drawn a great deal of interest recently, as they provide enhanced mechanical or rheological properties as well as new functionalities resulting from the various introduced inorganic species, which include silica particles, quantum dots (QDs), carbon nanotubes, exfoliated graphenes, *etc.* It is very interesting to note that for most of the hybrid hydrogels reported, the inorganic species were mainly introduced by supramolecular means, *i.e.*

hydrophobic interactions and host–guest interactions between the organic matrix and the inorganic moiety. A pioneering work in this newly emerging area was reported by Harada and coworkers.[72] This study utilized host–guest interaction between β-CDs of Py-β-CD/SWNT hybrids and dodecyl groups of poly(acrylic acid) to form supramolecular single walled carbon nanotubes (SWNT) hydrogel. This hybrid hydrogel with homogeneously dispersed SWNTs showed a chemical-responsive property, *i.e.* a "gel–sol" transition took place when a competitive host α-CD or guest ADA carboxylate (ADA-COONa) was mixed with the hydrogel ADACOONa interacts more strongly with β-CDs compared with dodecyl groups, inducing dissociation of complexes between β-CDs of Py-β-CD/SWNT hybrids and dodecyl groups of poly(acrylic acid). SWNTs were also introduced to α-CD/PEG-based hydrogel by Wang and Chen.[73] Here SWNTs were coated with PPG blocks of Pluronic® copolymers (PEG-*b*-PPO-*b*-PEG) and stabilized by the PEG blocks. Mixing these modified SWNTs with α-CD accelerated the formation of hydrogel based on α-CD/PEG. To some extent, it was unexpected that the viscosity and storage modulus of hybrid hydrogels would decline with an increase of SWNTs. Similarly, graphene sheets coated by the Pluronic® copolymer were employed to construct a hybrid hydrogel when they were mixed with α-CD.[74] The hybrid hydrogel containing graphene sheets also possesses lower viscosity and strength compared with the native ones. A study by Jiang *et al.* demonstrated ferrocene (Fc) HIC hydrogel structure with β-CD-modified quantum dots (β-CD@QD) as the core and Fc-ended diblock co-polymer of poly(*N*,*N*-dimethylacrylamide)-*b*-poly(*N*-isopropylacrylamide) (PDMA-*b*-PNIPAM) as the shell, to achieve an electrochemically active hydrogel at elevated temperatures.[71] Posessing two independent cross-linking mechanisms in the network structure, *i.e.,* the interchain aggregation of PNIPAM and inclusion complexation between CD and Fc on the surface of the quantum dots, the hydrogel undergoes sol–gel transition in response to temperature, addition of either an oxidizing agent or a competitive guest to Fc. In a similar recent study, Jiang *et al.* introduced a new type of hybrid hydrogel based on the "supracrosslink" concept composed of CdS quantum dots (Cds QDs) and a block copolymer of PDMA-*b*-PNIPAM.[75] A supramolecular structure HIC is formed *via* molecular recognition between β-CD@QD and azobenzene (AZO) end-functionalized block copolymer PDMA-*b*-PNIPAM (AZO-(PDMA-*b*-PNIPAM). In this HIC, PNIPAM formed the outer layer of the shell. After heating to the lower critical solution temperature (LCST) of PNIPAM, the PNIPAM block began to aggregate and finally led to hydrogels formed by the HIC. Thus the hydrogel had two distinctive crosslinks, the hydrophobic domains of PNIPAM and the multivalent nanoparticles. The PPR hydrogel, since the azo unit possesses a much greater ability to form ICs with α-CD than the PEG unit did, the threaded CD molecules were gradually pulled off leaving the PEG molecules as free chains. *cis*-Azo cannot bind to α-CD, so UV light irradiation to the solution made the azo–CD complex dissociate and then α-CD molecules returned to the PEG chains, which reconstructed the hydrogel. This study proved that the strength of the interactions is in the sequence of *trans*-azo-N^+/α-CD 4 PEG/α-CD 4 *cis*-azo-N^+/α-CD.

9.5 Nature of Inclusion Complex Formation

There are reports on inclusion complexation of pseudo block copolymers of star-shaped polymers. Since CD normally has 6–8 primary hydroxy groups on the upper rim of the cavity, it is convenient to initiate polymerization and form star-shaped polymers with several arms. Li and coworkers designed and synthesized a β-CD core four-arm PNIPAM and functionalized PEGs with ADA groups at one or both of its ends.[77] Thus they obtained non-covalently linked block copolymers with different architectures by inclusion complexation. The thermo-sensitive behavior of the β-CD-core star PNIPAM in the block copolymers was changed significantly, *i.e.* the LCST of these self-assembling systems was greatly increased depending on the ratio of the ADA moiety to the β-CD core and/or the length of the PEG blocks. In designing and fabricating new supramolecular materials, reversibility of the system provides a superior advantage to conventional hydrogels. Material reversibility can be realized by reversible non-covalent bonds. Two linear polymers comprising respective host (*e.g.* β-CD) and guest (*e.g.* ADA) as side groups form networks in water.[78] Two multi-arm star polymers with respective end groups of β-CD and cholesterol derivatives connect to each other forming a hydrogel due to the interaction between β-CD and cholesterol moieties.[79] Bridged CDs (two CDs covalently linked *via* a very short chain) were proven to be able to cross-link linear random copolymers of NIPAM and ADA-containing units, leading to gel formation.[80] Such bridged CD caused a remarkable LCST decrease of the copolymer from 35 °C to around 15 °C as a result of the restriction to the mobility and solubility of the polymer. γ-CD can include double-strand linear poly(ethylene imine) (LPEI) because of its wide cavity. When γ-CD was mixed with (PEO-*b*-LPEI)-g-dextran, a double-strand complex of γ-CD and LPEI grafts formed at pH 10 leading to network formation. When the pH value was adjusted to 4, LPEI was protonated, thus, γ-CD could not remain threaded on the LPEI chain, leading to an obvious viscosity decrease.[81] Free CDs can form a CD tube after crosslinking reaction by epichlorohydrin.[82] Then the resultant CD tube can accommodate two long alkyl chains from both its ends serving as a crosslinker. Therefore, a polymeric network was formed by addition of the α-CD tubes to the solution of PEG monocetylether-g-dextran.[83] It is well known that great contributions to supramolecular hydrogels came from CD-based PPR, using the microcrystalline area of CDs as a physical crosslink. After the discovery of the "molecular necklace" formed by PEG threading into a series of α-CDs,[30] Li *et al.* reported the first "sol–gel" transition of α-CD and PEG in aqueous solution in 1994.[31] This discovery initiated broad interest and active research which has continuously progressed in the past fifteen years. Many different methods of forming such CD-based gels and their associated architecture have since been reported (Figure 9.9).

The cross section of the polymers is correlated with the CD cavity size.[34,35,46,84] The binding complex between host and guest depends on the host cavity size and properties of the guests. With a larger cavity than α-CD and β-CD, γ-CD can include two LPEIs. When γ-CD was mixed with polysaccharides with

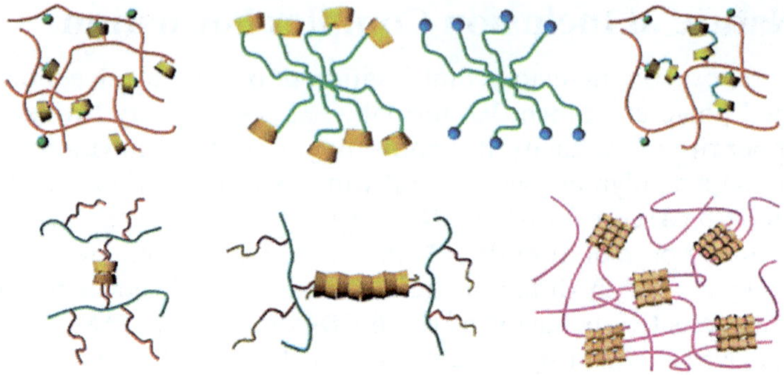

Figure 9.9 Supramolecular structures formed by inclusion complex crosslinks. Reproduced from ref. 18 with permission from the Royal Society of Chemistry.

grafts of block copolymer PEG-*b*-LPEI, a network structure formed, induced by the double-strand complex of γ-CD and LPEI grafts at pH 10.[81]

9.5.1 Inclusion Complex Driven by Micellization

The use of amphiphilic macromolecules as building blocks has drawn increasing interest because their controllable self-assembly and morphology transformation of the assemblies can be realized by tuning the amphiphilicity.[30] Generally, once part of a macromolecule containing a guest moiety is connected to CDs *via* inclusion complexation, the part will become more hydrophilic and thus the amphiphilicity of the macromolecule as a whole is turned. The location and distribution of the non-covalent binding sites along the polymer chains affect the overall structure. Cho and Allcock demonstrated that a block copolymer composed of two hydrophobic blocks, *i.e.* polystyrene and adamantyl polyphosphazene, could be converted to an amphiphilic copolymer when the adamantyl units enter the cavities of the added β-CDs. Thus micelles with polystyrene as the core and β-CD-modified polyphosphazene as the shell were induced.[85] Modification of polymer chains with CDs could also allow for the disassembly of micelles of amphiphilic copolymers. For example, alternating copolymers composed of sodium maleate and dodecyl vinyl ether units initially formed micelles in water due to the hydrophobic interactions between the dodecyl chains. When free α-CD was introduced, the inclusion complexation of dodecyl with α-CD competed with and overcame the existing hydrophobic interactions and then induced the dissociation of the micelles.[86]

9.6 Stimuli-Responsive Features of Hydrogels

The non-covalent nature of supramolecular hydrogels allows tunable response of the soft material. Many reports on CD-based hydrogels are termed "smart" polymers because they can be easily regulated by external stimuli. CD molecular recognition of a specific polymeric moiety drives the interaction in resulting hydrogel formation by responding to environmental changes, such as mechanical stress, temperature, pH, light, or chemical reagent. The tunable and efficient responsiveness of hydrogels depends on their network structure and crosslink strategies. Comprised of non-covalent bonds, inclusion complexation is a suitable method to serve as a reversible crosslinking factor due to its competitive nature. External stimuli such as light, pH, and temperature can regulate the mobility or shape of PPR formation, hence the overall physicochemical properties of the solution, due to introduction of stimuli-responsive moieties at the terminals of the PEG chain. In this section, we will present recent progress in stimuli-responsive and even reversible hydrogels that use ICs as crosslinks.

9.6.1 Mechanical

The most significant character of a supramolecular hydrogel is its shear-thinning behavior. Shear-thinning hydrogels and their potential use as injectable drug delivery systems have been proposed.[12] Recently, Chee *et al.* studied the release of proteins from such supramolecular hydrogels.[87] They reported erosion of PEO-α-CD PPR hydrogel to be the dominant factor for release of the proteins. Zhao *et al.* studied the shear-thinning properties of amphiphilic poly(ε-caprolactone)-poly(ethylene glycol)-poly(ε-caprolactone) (PCL-PEG-PCL) block copolymers and inclusion complexation with α-CD.[12] The storage modulus of these hydrogels is greater than the loss modulus over the entire range of frequency, and their apparent viscosity decreases with the increase of shear rate. In addition, their rheological properties could be tailored by changing the molecular weight of the PEG component in the block copolymers, the concentration of the block copolymer, and the actual molar ratio of the block copolymer and α-CD, which offers a means of controlling the mechanical properties for particular biomedical applications. Moreover, it was observed that a disrupted sol phase could be turned reversibly into a gel after shearing and then standing for a particular period of time. This property is important for an injectable drug delivery matrix through fine needles. People are fascinated by living nature's faculty to spontaneously repair damage. One unique property of supramolecular hydrogels is the ability to self-heal, mimicking the human tissue operational capability to restore. Recently, Harada and coworkers reported good examples of realizing such self-healing behavior of supramolecular hydrogels on a macroscopic scale.[88,89] The two cut pieces could be rejoined and the crack was sufficiently healed to form one gel. Karim and Loh reported that a PLLA/DMAEMA/PEGMA and α-CD hydrogel demonstrated self-healing properties.[39] The additive weak non-covalent interactions present in the network allow for associations strong enough to form a gel, yet weak enough for the network to be disrupted when exposed to external stimuli such as mechanical stress (Figure 9.10).

These mechanical properties of supramolecular hydrogels provide potential as injectable hydrogels.

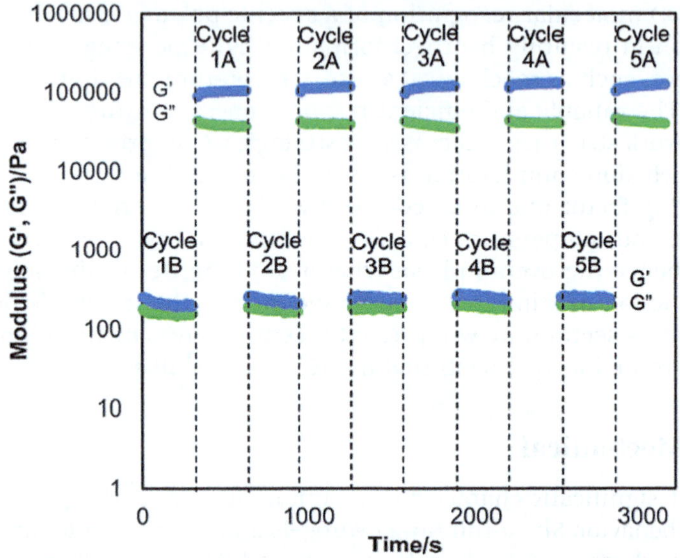

Figure 9.10 The PLED2 (10% w/v)/α-CD (10% w/v) hydrogel subjected to a self-healing cycle amplitude sweep test at 2 strain points, (A) 0.1% and (B) 50%, with 300 s at each strain point, at 37 °C. Reproduced from ref. 39 with permission from the Royal Society of Chemistry.

9.6.2 Temperature

Temperature is one of the extensively studied environmental triggers, which is crucial to physiological applications of hydrogels. There has been a large amount of research work on chemical hydrogels responsive to temperature based on temperature-sensitive polymers, particularly PNIPAM.[90–92] In general, complexation between host and guest molecules induced the overall properties of the hydrogel. For example, upon heating, in the hydrogel of α-CD and PEG, the α-CD rings will slide off, leading to the dissociation of the hydrogel. Based on this principle, a significant amount of work on the temperature responses of PPR hydrogels has been reported with various polymers such as block copolymer,[93] brush polymer,[27] super-branched polymer[94] *etc.* The hydrogel itself is thermo-reversible. Heating the PEG/α-CD hydrogels results in a homogeneous solution, and the sol returns to gel again after cooling. Besides PPR hydrogels, some of the other host–guest hydrogels with very different structures share a similar responsiveness to temperature. Hydrogels made of an 8-arm star polymer crosslinked by the association of β-CD and ADA[69,70] exhibited satisfactory thermo-reversibility upon heating and cooling steps, although the pair has a high binding constant. In the dynamic rheology measurement, as temperature increased from 4 °C to 37 °C, a dramatic decrease of *G′* and *G″* was observed. Below 17 °C, *G′* was larger than *G″*, showing the solid character of soft material. The curves of *G′* and *G″* intercepted each other at a temperature of 17 °C, indicating the gel-to-sol transition. The heating–cooling cycles were repeated several times without an apparent modulus loss. Similar results were observed in molecular recognition hydrogels with a different polymeric backbone and even higher binding ability.[5] These results demonstrated that the inherited thermo-reversible nature of hydrogels made of molecular recognition is the basic character of this type of materials.

9.6.3 UV and Visible Light Irradiations

Light-responsive systems have attracted growing interest since they are a clean trigger source and can be tuned remotely. In such studies, guest molecules of azobenzene and its derivatives have played a major role as their *cis* and *trans* forms possess very different abilities in complexation with CD hosts, and the *trans–cis* isomerization can be realized by UV light irradiation. Linear polymers modified with azo groups on their side chains were used more often to achieve light responsiveness.[95,96] Harada and his coworkers reported several photo-responsive hydrogels, most of their sol–gel transitions controlled by the reversible binding between CD and azobenzene.[97] For example, by mixing a CD-functionalized curdlan with azo-modified poly(acrylic acid),[98] the CD–azo binding served as crosslinker; thus the *trans*-azo induced gel state was converted to sol under UV irradiation. Azobenzene is a well-known photosensitive molecule which transforms from the *trans* isomer to the *cis* isomer upon UV irradiation, so it is possible to regulate inclusion complexation between azo and α-CD.[99] The azo moiety is captured in the cavity of α-CD when azo forms the *trans* isomer but not in the *cis* isomer due to steric hindrance (Figure 9.11).[95]

A review by Chen and coworkers explained that addition of a water-soluble azo compound to PPR hydrogel of PEG and α-CD in water removed the α-CD rings from the PEG chains resulting in a transparent solution, due to the stronger complexation of *trans*-azo/α-CD than that of PEG/α-CD. Subsequent irradiation with UV light drove the α-CD cavities to move back to the PEG chains since

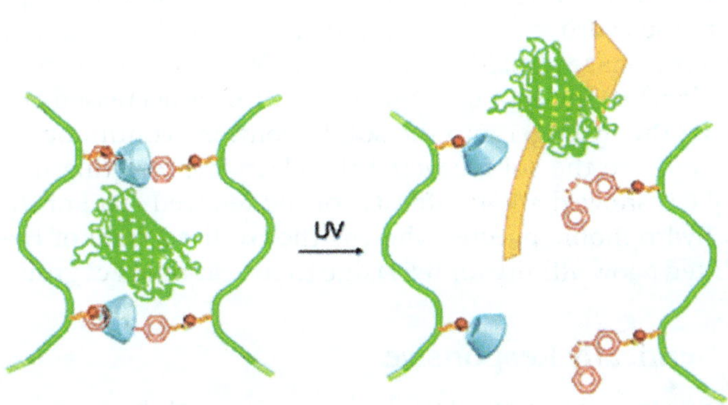

Figure 9.11 Schematic illustration of stimuli-responsible gate for pseudopolyrotaxane formation between α-CD and PEG. Reproduced from ref. 95 with permission from the Royal Society of Chemistry.

UV light induced isomerization of *trans*-azo to its *cis* form, and consequently lost the ability to form a complex with α-CD, resulting in a hydrogel again. Further irradiation with visible light can make a gel-to-sol transition occur and start another cycle. For PPR hydrogels composed of PEG and α-CDs, although both components are not active for light irradiation, photoinduced reversible gel-to-sol transitions were realized by pure "supramolecular routes" in Jiang's group.[76] Based on the competitive host–guest interactions, this reversible sol–gel process was observed repeatedly by alternation of UV and visible light irradiation. The same principle was also successfully employed in HIC hydrogels, where the mechanical strength was greatly enhanced by clay.[41]

9.6.4 Electrical and Redox Stimuli

Redox-responsive systems may be applied to benign electrofunctional systems, such as electrically switchable materials.[100,101] Inclusion complexation between CD and Fc is well-known and has been extensively studied because of its reversible association controlled by the redox state of Fc, which can be tuned either electrochemically or chemically, *i.e.* with redox reagents. A redox-responsible hydrogel system was achieved by a combination of β-CD, PAA with dodecyl chains attached randomly, and a redox-responsive guest, ferrocene-carboxylic acid (Fc-A).[102] The initial hydrogel of the PAA formed due to the hydrophobic interactions of the dodecyl chains. But the hydrogel was dissociated by addition of β-CD which formed an IC with the alkyl chains. The subsequent addition of guest Fc-A led to the transfer of β-CD from the alkyl chains to Fc-A, thus the ternary mixture exhibited the gel behavior again due to the recovered hydrophobic interactions of the dodecyl chains. When Fc-A was tuned to its oxidized state, it lost its ability to form a complex, β-CD formed a complex with the dodecyl chains again and consequently the mixture became a sol. The HIC hydrogels containing Fc-ended block copolymer mentioned earlier (Fc–HIC) can also achieve similar responses,[71] *i.e.* the gel-to-sol transition took place after addition of an oxidizing agent, *e.g.* $K_3Fe(CN)_6$ to the hydrogel. Nakahata *et al.* investigated the effect of the redox reagents on the phase transition of the pAA-6βCD/pAA-Fc hydrogel.[88] Adding NaClO (oxidant) to the pAA-6βCD/pAA-Fc hydrogel decreased the viscosity, transforming the hydrogel into the sol. In contrast, continuous addition of GSH (reductant) to the sol recovered the elasticity, reverting it back to the hydrogel. β-CD showed a high affinity for the reduced state of the Fc group due to its hydrophobic nature, whereas the oxidized state of the Fc group (Fc^+) exhibited a low affinity for β-CD due to the cationic Fc^+ group.

9.6.5 Chemically Responsive

The chemical responsiveness of the hydrogels is mostly based on the dynamic and competitive nature of molecular recognition.[72] For example, the hydrogel composed of trinitrobenzene (TNB) modified β-CD can first form a supramolecular polymer, then hydrogels at certain concentrations.[103,104] The hydrogel turned into sol by addition of either ADA-containing molecules as a competitive guest or urea as a denaturing reagent, which could break the hydrogen bonds between CDs. Meanwhile, the hydrogel was induced to sol by an excess of native β-CD as a competitive host. In addition, the gel-to-sol transition also occurred upon addition of methyl orange (MO) as a competitive guest, followed by sol-to-gel transition upon addition of α-CD, which was a stronger host for MO. This reversible gel-to-sol transition was performed several times upon alternate additions of MO and α-CD. This perfect chemical reversibility can be performed due to the ultra-low solid content of hydrogels made of the supramolecular polymer.

9.7 Conclusion

In this chapter, we presented a comprehensive summary of hydrogels formed from inclusion complexation of CD and polymers. α-CD/PEG-based hydrogels are the most common material being studied owing to their strong affinity towards each other. Modification of PEG-based polymers and the availability of other monomers led to the development of other CD-based hydrogels. The major driving forces for forming CD ICs are hydrophobic and van der Waals interactions between the inner surface of the CD ring and the hydrophobic sites of the guests. Numerous studies have been reported on other monomer gels which incorporate CDs to form ICs. In all, considering the youth of both polymer and supramolecular chemistry, and all the achievements already realized, their combination will certainly bring us novel materials with unprecedented properties in the near future. Therefore, besides the need for further fundamental research in this field, future research should also aim to develop applications based on these versatile materials, which are especially promising in biomedicine. Finally, the introduction of stimuli-responsive supramolecular guests along the polymer side-chain or responsive supramolecular hosts is allowing development of multi-stimuli responsive systems. These systems will furnish new molecular logic-gates, highly sensitive detection and diagnosis for lab-on-a-chip technologies, or drug/gene delivery carriers with improved targeted release. The hydrophobic nature of most potent active pharmaceutical ingredients, together with the inherent capability of supramolecular hosts to complex with hydrophobic guests, will definitely open the way to new therapies benefiting from these responsive systems.

References

1. A. Harada, A. Hashidzume, H. Yamaguchi and Y. Takashima, *Chem. Rev.*, 2009, **109**, 5974–6023.
2. H.-G. Wu, X.-J. Ju, R. Xie, Y.-M. Liu, J.-G. Deng, C. H. Niu and L.-Y. Chu, *Polym. Adv. Technol.*, 2011, **22**, 1389–1394.
3. P. Mi, X.-J. Ju, R. Xie, H.-G. Wu, J. Ma and L.-Y. Chu, *Polymer*, 2010, **51**, 1648–1653.

4. H. Yamaguchi, Y. Kobayashi, R. Kobayashi, Y. Takashima, A. Hashidzume and A. Harada, *Nat. Commun.*, 2012, **3**, 603.
5. E. A. Appel, F. Biedermann, U. Rauwald, S. T. Jones, J. M. Zayed and O. A. Scherman, *J. Am. Chem. Soc.*, 2010, **132**, 14251–14260.
6. K. Y. Lee and D. J. Mooney, *Chem. Rev.*, 2001, **101**, 1869–1879.
7. A. S. Hoffman, *Adv. Drug Delivery Rev.*, 2002, **54**, 3–12.
8. T. R. Hoare and D. S. Kohane, *Polymer*, 2008, **49**, 1993–2007.
9. J. S. Boateng, K. H. Matthews, H. N. E. Stevens and G. M. Eccleston, *J. Pharm. Sci.*, 2008, **97**, 2892–2923.
10. T. Vermonden, R. Censi and W. E. Hennink, *Chem. Rev.*, 2012, **112**, 2853–2888.
11. M. K. Nguyen and E. Alsberg, *Prog. Polym. Sci.*, 2014, **39**, 1235–1265.
12. S.-P. Zhao, L.-M. Zhang and D. Ma, *J. Phys. Chem. B*, 2006, **110**, 12225–12229.
13. I. Terekhova, R. De Lisi, G. Lazzara, S. Milioto and N. Muratore, *J. Therm. Anal. Calorim.*, 2008, **92**, 285–290.
14. M. V. Rekharsky and Y. Inoue, *Chem. Rev.*, 1998, **98**, 1875–1918.
15. R. De Lisi, G. Lazzara, S. Milioto and N. Muratore, *Chemosphere*, 2007, **69**, 1703–1712.
16. A. Harada, Y. Takashima and M. Nakahata, *Acc. Chem. Res.*, 2014, **47**, 2128–2140.
17. A. Harada, Y. Takashima and H. Yamaguchi, *Chem. Soc. Rev.*, 2009, **38**, 875–882.
18. G. Chen and M. Jiang, *Chem. Soc. Rev.*, 2011, **40**, 2254–2266.
19. X. Liao, G. Chen and M. Jiang, *Polym. Chem.*, 2013, **4**, 1733–1745.
20. F. Yuen and K. C. Tam, *Soft Matter*, 2010, **6**, 4613–4630.
21. K. L. Liu, Z. Zhang and J. Li, *Soft Matter*, 2011, **7**, 11290–11297.
22. J. Li, X. Ni and K. W. Leong, *J. Biomed. Mater. Res., Part A*, 2003, **65**, 196–202.
23. J. Li, X. Li, Z. Zhou, X. Ni and K. W. Leong, *Macromolecules*, 2001, **34**, 7236–7237.
24. L. Jongpaiboonkit, Z. Zhou, X. Ni, Y. Z. Wang and J. Li, *J. Biomater. Sci., Polym. Ed.*, 2006, **17**, 747–763.
25. K. M. Huh, T. Ooya, W. K. Lee, S. Sasaki, I. C. Kwon, S. Y. Jeong and N. Yui, *Macromolecules*, 2001, **34**, 8657–8662.
26. K. M. Huh, Y. W. Cho, H. Chung, I. C. Kwon, S. Y. Jeong, T. Ooya, W. K. Lee, S. Sasaki and N. Yui, *Macromol. Biosci.*, 2004, **4**, 92–99.
27. L. He, J. Huang, Y. Chen, X. Xu and L. Liu, *Macromolecules*, 2005, **38**, 3845–3851.
28. E. Sabadini and T. Cosgrove, *Langmuir*, 2003, **19**, 9680–9683.
29. T. Hoshino, M. Miyauchi, Y. Kawaguchi, H. Yamaguchi and A. Harada, *J. Am. Chem. Soc.*, 2000, **122**, 9876–9877.
30. A. Harada and M. Kamachi, *Macromolecules*, 1990, **23**, 2821–2823.
31. A. Harada, J. Li and M. Kamachi, *Macromolecules*, 1993, **26**, 5698–5703.
32. A. Harada, J. Li and M. Kamachi, *Macromolecules*, 1994, **27**, 4538–4543.
33. J. Li, A. Harada and M. Kamachi, *Polym. J.*, 1994, **26**, 1019–1026.

34. A. Harada, *Carbohydr. Polym.*, 1997, **34**, 183–188.
35. A. Harada, *Coord. Chem. Rev.*, 1996, **148**, 115–133.
36. A. Harada, J. Li and M. Kamachi, *Nature*, 1994, **370**, 126–128.
37. H. Ye, C. Owh and X. J. Loh, *RSC Adv.*, 2015, **5**, 48720–48728.
38. D. Kai, Z. W. Low, S. S. Liow, A. Abdul Karim, H. Ye, G. Jin, K. Li and X. J. Loh, *ACS Sustainable Chem. Eng.*, 2015, **3**, 2160–2169.
39. A. Abdul Karim and X. J. Loh, *Soft Matter*, 2015, **11**, 5425–5434.
40. M. Guo, M. Jiang, S. Pispas, W. Yu and C. Zhou, *Macromolecules*, 2008, **41**, 9744–9749.
41. X. Liao, G. Chen and M. Jiang, *Langmuir*, 2011, **27**, 12650–12656.
42. Z. Tian, C. Chen and H. R. Allcock, *Macromolecules*, 2013, **46**, 2715–2724.
43. F. Li, J. He, M. Zhang and P. Ni, *Polym. Chem.*, 2015, **6**, 5009–5014.
44. K. Huang, L. Li, J. Wang, Z. Zhou and X. Guo, *Colloid Polym. Sci.*, 2015, 1–9, DOI: 10.1007/s00396-015-3788-0.
45. L. Y. Tao Kong, A.-Y. Zhang and Z.-G. Feng, *Beilstein J. Org. Chem.*, 2014, **10**, 2461–2469.
46. A. Harada, M. Okada and M. Kamachi, *Bull. Chem. Soc. Jpn.*, 1998, **71**, 535–542.
47. A. Harada and M. Kamachi, *J. Chem. Soc., Chem. Commun.*, 1990, 1322–1323, DOI: 10.1039/C39900001322.
48. A. Harada, M. Okada, J. Li and M. Kamachi, *Macromolecules*, 1995, **28**, 8406–8411.
49. A. Harada, J. Li and M. Kamachi, *Chem. Lett.*, 1993, **22**, 237–240.
50. K. A. Udachin, L. D. Wilson and J. A. Ripmeester, *J. Am. Chem. Soc.*, 2000, **122**, 12375–12376.
51. H. Saito, H. Yonemura, H. Nakamura and T. Matsuo, *Chem. Lett.*, 1990, **19**, 535–538.
52. M. Watanabe, H. Nakamura and T. Matsuo, *Bull. Chem. Soc. Jpn.*, 1992, **65**, 164–169.
53. A. Harada, H. Adachi, Y. Kawaguchi, M. Okada and M. Kamachi, *Polym. J.*, 1996, **28**, 159–163.
54. L. P. Meier, M. Heule, W. R. Caseri, R. A. Shelden, U. W. Suter, G. Wenz and B. Keller, *Macromolecules*, 1996, **29**, 718–723.
55. A. Nelson, J. M. Belitsky, S. Vidal, C. S. Joiner, L. G. Baum and J. F. Stoddart, *J. Am. Chem. Soc.*, 2004, **126**, 11914–11922.
56. G. Wenz and B. Keller, *Macromol. Symp.*, 1994, **87**, 11–16.
57. W. Herrmann, B. Keller and G. Wenz, *Macromolecules*, 1997, **30**, 4966–4972.
58. G. Wenz, C. Gruber, B. Keller, C. Schilli, T. Albuzat and A. Müller, *Macromolecules*, 2006, **39**, 8021–8026.
59. Y. Y. Yoshinori Takashima, M. Otsubo, H. Yamaguchi and A. Harada, *Beilstein J. Org. Chem.*, 2012, **8**, 1594–1600.
60. A. Harada, J. Li, S. Suzuki and M. Kamachi, *Macromolecules*, 1993, **26**, 5267–5268.
61. A. Harada, T. Nishiyama, Y. Kawaguchi, M. Okada and M. Kamachi, *Macromolecules*, 1997, **30**, 7115–7118.

62. A. Harada, Y. Kawaguchi, T. Nishiyama and M. Kamachi, *Macromol. Rapid Commun.*, 1997, **18**, 535–539.
63. L. Huang, E. Allen and A. E. Tonelli, *Polymer*, 1998, **39**, 4857–4865.
64. Y. Kawaguchi, T. Nishiyama, M. Okada, M. Kamachi and A. Harada, *Macromolecules*, 2000, **33**, 4472–4477.
65. X. Shuai, F. E. Porbeni, M. Wei, T. Bullions and A. E. Tonelli, *Macromolecules*, 2002, **35**, 3778–3780.
66. X. Shuai, F. E. Porbeni, M. Wei, T. Bullions and A. E. Tonelli, *Macromolecules*, 2002, **35**, 2401–2405.
67. A. Harada, H. Okumura, M. Okada, S. Suzuki and M. Kamachi, *Chem. Lett.*, 2000, **29**, 548–549.
68. S. Osman, G. M. Soliman, M. Amin and A. Zaky, *Int. J. Pharma. Pharma. Sci.*, 2014, **6**, 591–597.
69. F. van de Manakker, M. van der Pot, T. Vermonden, C. F. van Nostrum and W. E. Hennink, *Macromolecules*, 2008, **41**, 1766–1773.
70. F. van de Manakker, L. M. J. Kroon-Batenburg, T. Vermonden, C. F. van Nostrum and W. E. Hennink, *Soft Matter*, 2010, **6**, 187–194.
71. P. Du, J. Liu, G. Chen and M. Jiang, *Langmuir*, 2011, **27**, 9602–9608.
72. T. Ogoshi, Y. Takashima, H. Yamaguchi and A. Harada, *J. Am. Chem. Soc.*, 2007, **129**, 4878–4879.
73. Z. Wang and Y. Chen, *Macromolecules*, 2007, **40**, 3402–3407.
74. S.-Z. Zu and B.-H. Han, *J. Phys. Chem. C*, 2009, **113**, 13651–13657.
75. J. Liu, G. Chen, M. Guo and M. Jiang, *Macromolecules*, 2010, **43**, 8086–8093.
76. X. Liao, G. Chen, X. Liu, W. Chen, F. Chen and M. Jiang, *Angew. Chem., Int. Ed.*, 2010, **49**, 4409–4413.
77. Z.-X. Zhang, X. Liu, F. J. Xu, X. J. Loh, E.-T. Kang, K.-G. Neoh and J. Li, *Macromolecules*, 2008, **41**, 5967–5970.
78. A. Charlot and R. Auzély-Velty, *Macromolecules*, 2007, **40**, 9555–9563.
79. F. van de Manakker, T. Vermonden, N. el Morabit, C. F. van Nostrum and W. E. Hennink, *Langmuir*, 2008, **24**, 12559–12567.
80. O. Kretschmann, S. W. Choi, M. Miyauchi, I. Tomatsu, A. Harada and H. Ritter, *Angew. Chem., Int. Ed.*, 2006, **45**, 4361–4365.
81. Y. K. Joung, T. Ooya, M. Yamaguchi and N. Yui, *Adv. Mater.*, 2007, **19**, 396–400.
82. A. Harada, J. Li and M. Kamachi, *Nature*, 1993, **364**, 516–518.
83. T. Ikeda, T. Ooya and N. Yui, *Macromol. Rapid Commun.*, 2000, **21**, 1257–1262.
84. A. Harada, M. Okada and M. Kamachi, *Acta Polym.*, 1995, **46**, 453–457.
85. S. Y. Cho and H. R. Allcock, *Macromolecules*, 2009, **42**, 4484–4490.
86. D. Taura, A. Hashidzume and A. Harada, *Macromol. Rapid Commun.*, 2007, **28**, 2306–2310.
87. P. L. Chee, A. Prasad, X. Fang, C. Owh, V. J. Yeo and X. J. Loh, *Mater. Sci. Eng. C*, 2014, **39**, 6–12.
88. M. Nakahata, Y. Takashima, H. Yamaguchi and A. Harada, *Nat. Commun.*, 2011, **2**, 511.

89. Y. Zheng, A. Hashidzume, Y. Takashima, H. Yamaguchi and A. Harada, *Nat. Commun.*, 2012, **3**, 831.
90. V. Bütün, S. Liu, J. V. M. Weaver, X. Bories-Azeau, Y. Cai and S. P. Armes, *React. Funct. Polym.*, 2006, **66**, 157–165.
91. H. Liu, Y. Zhang, J. Hu, C. Li and S. Liu, *Macromol. Chem. Phys.*, 2009, **210**, 2125–2137.
92. B.-W. Liu, H. Zhou, S.-T. Zhou, H.-J. Zhang, A.-C. Feng, C.-M. Jian, J. Hu, W.-P. Gao and J.-Y. Yuan, *Macromolecules*, 2014, **47**, 2938–2946.
93. K. Sui, X. Shan, S. Gao, Y. Xia, Q. Zheng and D. Xie, *J. Polym. Sci., Part A: Polym. Chem.*, 2010, **48**, 2143–2153.
94. X. Zhu, L. Chen, D. Yan, Q. Chen, Y. Yao, Y. Xiao, J. Hou and J. Li, *Langmuir*, 2004, **20**, 484–490.
95. K. Peng, I. Tomatsu and A. Kros, *Chem. Commun.*, 2010, **46**, 4094–4096.
96. Y.-L. Zhao and J. F. Stoddart, *Langmuir*, 2009, **25**, 8442–8446.
97. I. Tomatsu, A. Hashidzume and A. Harada, *Macromolecules*, 2005, **38**, 5223–5227.
98. S. Tamesue, Y. Takashima, H. Yamaguchi, S. Shinkai and A. Harada, *Angew. Chem., Int. Ed.*, 2010, **49**, 7461–7464.
99. T. Ikeda, T. Ooya and N. Yui, *Polym. J.*, 1999, **31**, 658–663.
100. Y. Liu, A. H. Flood, P. A. Bonvallet, S. A. Vignon, B. H. Northrop, H. R. Tseng, J. O. Jeppesen, T. J. Huang, B. Brough, M. Baller, S. Magonov, S. D. Solares, W. A. Goddard, C. M. Ho and J. F. Stoddart, *J. Am. Chem. Soc.*, 2005, **127**, 9745–9759.
101. C. A. Nijhuis, F. Yu, W. Knoll, J. Huskens and D. N. Reinhoudt, *Langmuir*, 2005, **21**, 7866–7876.
102. I. Tomatsu, A. Hashidzume and A. Harada, *Macromol. Rapid Commun.*, 2006, **27**, 238–241.
103. W. Deng, H. Yamaguchi, Y. Takashima and A. Harada, *Chem.–Asian J.*, 2008, **3**, 687–695.
104. W. Deng, H. Yamaguchi, Y. Takashima and A. Harada, *Angew. Chem., Int. Ed.*, 2007, **46**, 5144–5147.

CHAPTER 10

Thermogelling Polymers: A Cutting Edge Rheology Modifier

SING SHY LIOW[a], QINGQING DOU[a], DAN KAI[a], ANIS ABDUL KARIM[a], KANGYI ZHANG[a], AND XIAN JUN LOH*[a,b]

[a]Institute of Materials Research and Engineering, A*STAR (Agency for Science, Technology and Research), 2 Fusionopolis Way, Innovis, #08-03, Singapore 138634, Singapore; [b]Department of Materials Science and Engineering, National University of Singapore, Singapore 117574, Singapore
*E-mail: lohxj@imre.a-star.edu.sg

10.1 Introduction

Hydrogels, owing to their high water content and highly tunable properties, are a key class of soft materials suitable for a variety of biomedical applications. These 3D polymeric networks can be formed by chemical or physical crosslinks. They can hold a large amount of water without dissolution. Hydrogels made from polymeric networks produced *via* the chemical crosslinking route are typically tough and elastic. These are properties which are highly desirable in dynamic environments such as cartilage, skin and cardio-related devices.[1,2] On the other hand, hydrogels derived from physically crosslinked polymeric networks (*e.g.* molecular self-assembly, hydrogen bonding, hydrophilic/hydrophobic interaction, host–guest inclusion complex),[3–6] formed by a simple phase transition (sol–gel) in water without any chemical reaction or external energy source, are particularly interesting because the systems allow simplicity and safety for applications.[7]

RSC Polymer Chemistry Series No. 20
Polymers for Personal Care Products and Cosmetics
Edited by Xian Jun Loh

Published by the Royal Society of Chemistry, www.rsc.org

A thermogelling polymer is a physical gel that undergoes a reversible sol–gel transition as the temperature is monotonically increased. Thermogelling polymers can be easily administered *via* injection and the subsequent *in situ* gelation.[8–12] A typical injectable thermogelling polymer system is formulated by simple mixing of therapeutics in a hydrogel below the gel transition temperature. After injection, sol-to-gel transition occurs under physiological temperature to transform the minimally viscous solution into a therapeutic conveyance depot. This method is advantageous because (i) it avoids invasive surgery for implantation; (ii) the high water content of the hydrogel improves compatibility with the injection site; (iii) sterilization is done easily by syringe filtration; (iv) peptides are encapsulated at low temperature, which prevents denaturation due to organic solvent interaction or high temperature dissolution; (v) the biodegradable thermogel can be excreted from the body after achieving its intended purpose; (vi) the rate of therapeutic release can be easily tailored by changing the formulation.[10]

The objective of this chapter is to review the various thermogels reported in recent literature with an emphasis on the synthetic procedures and polymeric assembly mechanisms, gel resorbability and applications for therapeutics delivery, cell encapsulation and tissue repair.

10.2 Synthesis and Self-Assembly of Thermogelling Polymers

Thermogels exhibit reversible sol–gel transition as the temperature is monotonically increased and precipitate or transform from gel to solution at elevated temperatures. This phase change behavior is reversible because the gel is formed by physical crosslinks between the polymer chains. Unlike permanent crosslinks formed by chemical reactions, physical crosslinks in thermoresponsive copolymers are formed *via* hydrophilic/hydrophobic physical associations. Thermoresponsive copolymers consist of hydrophilic and hydrophobic segments, which can self-assemble into polymeric micelles in water. The hydrophobic segments form the core of the micelles while the hydrophilic chains interact with water molecules at the corona.

Thermoresponsive copolymers can be synthesized *via* different methods including ring opening polymerization, atom transfer radical polymerization (ATRP), reversible addition–fragmentation chain transfer (RAFT) polymerization and polyurethane formation (polycondensation). Each method aims to produce amphiphilic copolymers that consist of hydrophilic and hydrophobic segments. As gel formation is mainly driven by hydrophobic attraction, fine-tuning the ratio of hydrophilic and hydrophobic segments is the key to achieving thermogelling properties. Poly(ethylene glycol) (PEG) and poly(propylene glycol) (PPG) are commonly used in thermogels because of their well-known biocompatibility. PEG has a lower critical solution temperature (LCST) in the range of 100–150 °C in water, while PPG has a LCST range of 10–30 °C in water. With PEG as the hydrophilic segment and PPG as the hydrophobic segment, this amphiphilic copolymer shows thermogelling behavior at physiological temperature. Besides PEG and PPG, typical polymers that exhibit a LCST include poly(*N*-isopropylacrylamide) (PNIPAAm), poly(vinyl ether) (PVE), poly(*N*,*N*-diethylacrylamide) (PDEAM), poly(*N*-vinyl alkyl amide) and poly(*N*-vinyl caprolactam), as listed in Table 10.1.[13] A comprehensive table showing various thermoresponsive polymers and their respective LCST and UCST is

Table 10.1 LCST of several typical thermoresponsive polymers.[13]

Polymer	Chemical structure	LCST (°C)
Poly(*N*-isopropylacrylamide) (PNIPAAm)	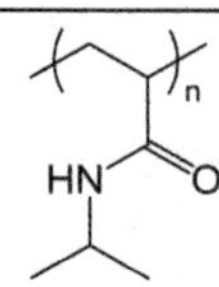	32
Poly(*N*,*N*-diethylacrylamide) (PDEAM)	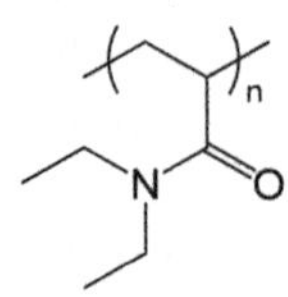	25
Poly(*N*-ethyl methacrylamide) (PNEMAM)	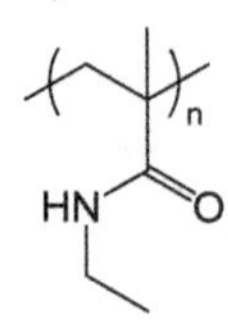	58
Poly(methyl vinyl ether) (PMVE)	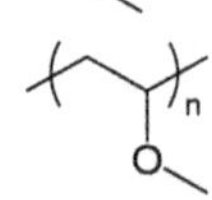	34
Poly(2-ethoxyethyl vinyl ether) (PEOVE)	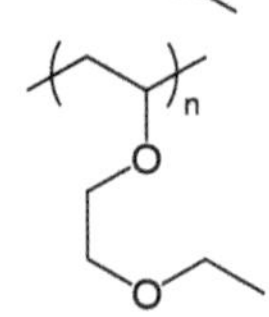	20
Poly(*N*-vinyl isobutyramide) (PNVIBAM)	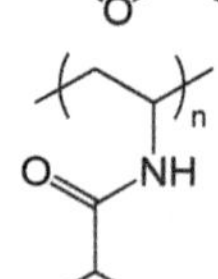	39
Poly(*N*-vinyl *n*-butyramide) (PNVBAM)	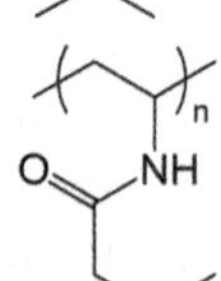	32
Poly(*N*-vinyl caprolactam) (PNVCa)	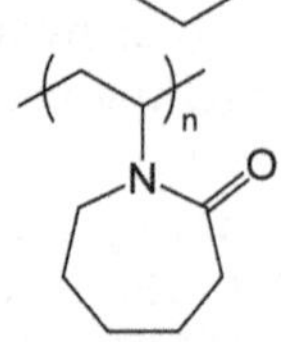	30–50

listed in ref. 14 and the interested reader is referred to a notable review[15] for additional explanation on this phase change behavior.

10.2.1 PEG-Based Block Copolymers

PEG–PPG–PEG triblock copolymers, also known as Pluronic® (BASF) or Poloxamer (ICI), consist of 30% PPG hydrophobic segment and 70% PEG hydrophilic segment. As these gels are eroded (but non-degradable) within a few days *in vivo*, long term therapeutic release is not feasible. Numerous studies have provided different solutions, including crosslinking,[16,17] grafting,[18] copolymerization[19] or substituting PPG with other polyesters such as poly(D,L-lactic acid-*co*-glycolic acid) (PLGA),[20] polycaprolactone (PCL)[21–23] and poly([*R*]-3-hydroxybutyrate) (PHB)[24] (Figure 10.1, **1–4**). ABA-type triblock copolymers are synthesized in a two-step reaction: ring opening polymerization of B block using methoxy-PEG as initiator, then condensation reaction to link two B blocks together using diisocyanate as coupling agent. ABA copolymers consisting of PLGA as the middle block significantly increased the gel duration up to a few weeks. As the hydrophobic PLGA chain length increased, the gelation temperature and gelation concentration decreased. This indicates that increasing hydrophobicity enhances thermodynamic interaction associated with gelation. Compared to PLGA, PCL is more hydrophobic. PEG–PCL–PEG forms a gel at a lower polymer concentration compared to PEG–PLGA–PEG. PLGA (G : L ratio 2 : 8) and PCL exhibit three and ten times more hydrophobicity, respectively, than PPG.[22] Further to this, the hydrophobicity of PHB is typically higher than most biodegradable polyesters. However, PEG–PHB–PEG can only form micelles but thermogelation is not achievable at any temperature.[24] This is possibly due to the imbalanced hydrophilic–hydrophobic ratio.

BAB-type triblock thermogels, especially PLGA–PEG–PLGA, have been studied intensively since 2000.[25–27] The PLGA–PEG–PLGA (1500–1000–1500) thermogel (commercially available as ReGel®) is used in release studies of proteins and conventional therapeutics.[27] The synthesis of BAB-type amphiphilic copolymers is easier compared to the ABA-type. It is prepared by ring opening polymerization of lactic acid and glycolic acid cyclic monomers, using PEG 1000 Da as the initiator and tin octoate as the catalyst. Compared to PEG–PLGA–PEG, the BAB-type consists of more hydrophobic segments. Sol–gel transition of PLGA–PEG–PLGA occurs at a lower temperature due to enhanced hydrophobic interaction. Furthermore, the alteration of end groups from hydrophilic hydroxyl terminals to hydrophobic alkyl chains significantly lowers the sol–gel transition temperature by 10 °C.[25,28] The gelation concentration also becomes lower, from 12 wt% to 2 wt% (Figure 10.1, **5a** and **5b**). An even more complex case for the end group effect on thermogelling properties was found in an amphiphilic block copolymer with an ionizable group, reported by the same research group.[29] These studies draw a clear conclusion: tuning the balance of hydrophilicity and hydrophobicity is the key to achieving thermogelation.

ABA type

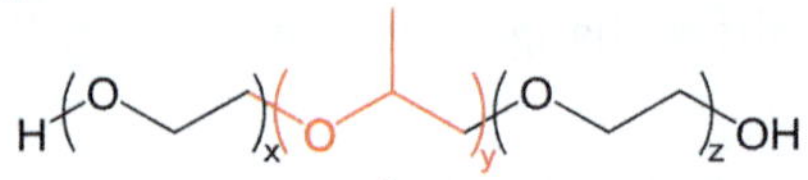

1

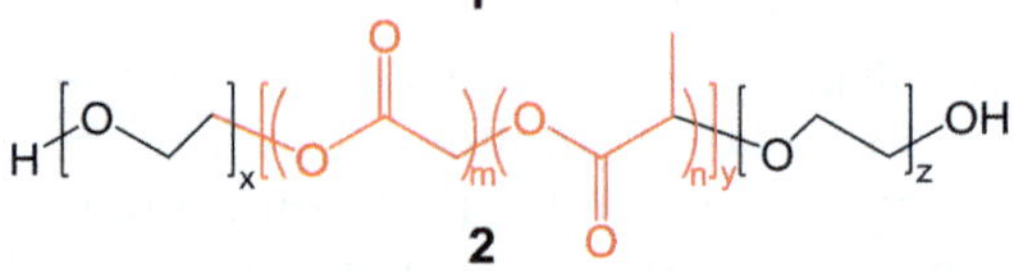

2

3

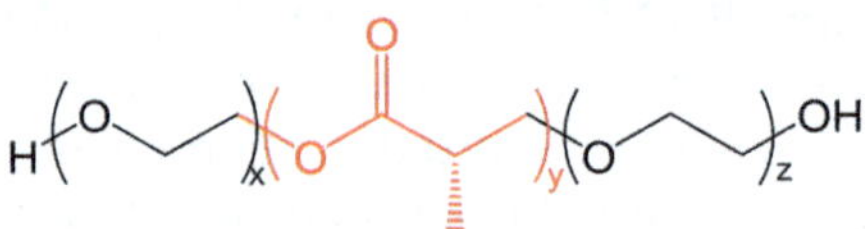

4

BAB type

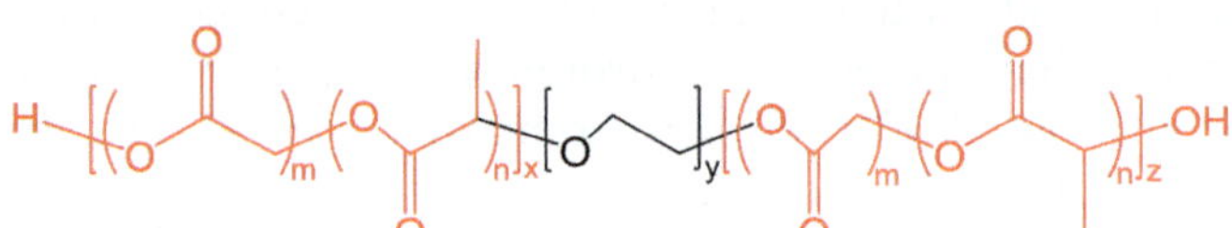

5

⇓

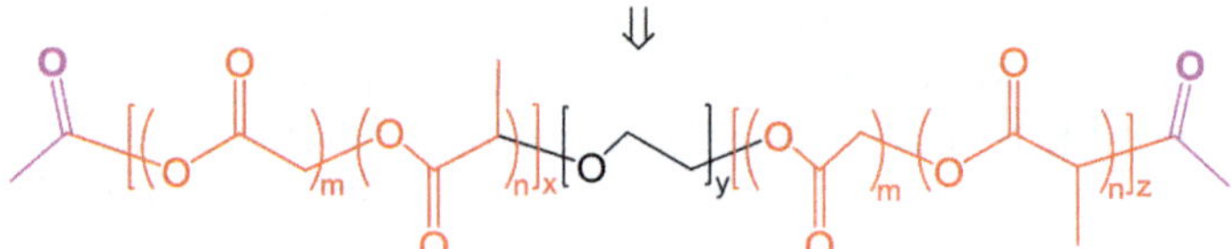

5a

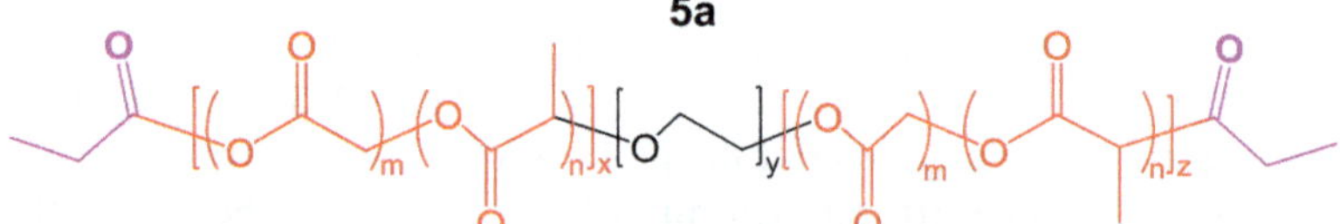

5b

Figure 10.1 Chemical structure of ABA-type and BAB-type triblock copolymers. (red: hydrophobic segments, pink: additional hydrophobic ends, blue: pH-responsive segments) **1**: Pluronic® PEG–PPG–PEG **2**: PEG–PLGA–PEG **3**: PEG–PCL–PEG **4**: PEG–PHB–PEG **5**: PLGA–PEG–PLGA **5a**: diacetate PLGA–PEG–PLGA **5b**: dipropionates PLGA–PEG–PLGA.

The physical gelation of PLGA–PEG–PLGA thermogel in water was studied using TEM, ^{13}C NMR and DLS;[25] the digital picture and schematic drawing of the micelle network are shown in Figure 10.2. The gelling mechanism is due to ordered packing of micelles. In the sol state, micelles are formed by self-assembly of amphiphilic block copolymers, where the hydrophobic PLGA and hydrophilic PEG form the core and the corona, respectively. At low temperatures, the aqueous solution is a suspension of micelles. As the temperature increases to the transition temperature, the micelles aggregate

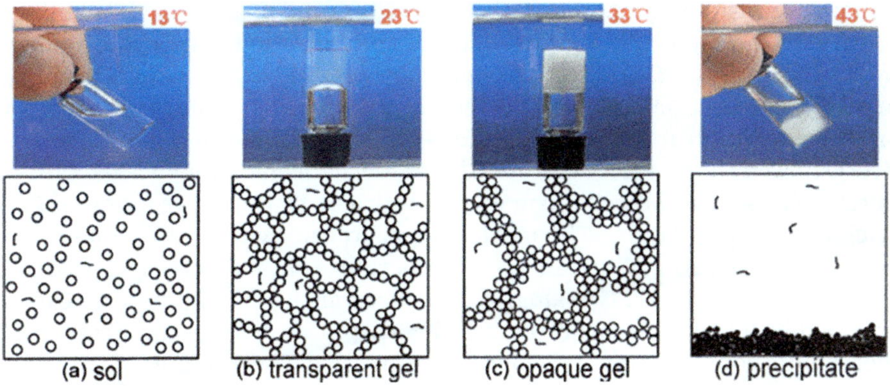

Figure 10.2 Visual observation (above) and schematic drawing of PLGA–PEG–PLGA micellar network showing thermogelling behavior (below (a)–(d).). Reproduced from Yu *et al.*[25] with permission from John Wiley & Sons © 2006 Wiley Periodicals, Inc.[25]

into a percolated micellar network due to hydrophobic interactions between them.[30] As the micelles aggregate into a coarser structure, with a mesh size in the order of the visible light wavelength, the gel becomes opaque. As the temperature continues to rise, excessive hydrophobicity destroys the micellar structure which leads to macroscopic precipitation.

Besides the type of hydrophobic block, block sequence (ABA and BAB), and end groups as mentioned above, thermogelling properties were also found to be adjusted by molecular weights and polydispersity indices of both hydrophilic and hydrophobic blocks.[31,32] Addition of salts (*e.g.* NaCl) can significantly tune the sol–gel transition temperature and critical gelation concentration (CGC) of triblock copolymers.[20] Mixing of two non-thermogellable copolymers can sometimes lead to thermogellation.[33,34]

Recently, studies on Pluronic® F-127 conjugation revealed more possibilities to achieve tunable properties. Shachaf *et al.* reported the synthesis of Pluronic® F-127 and fibrinogen crosslinked hydrogel, prepared by photopolymerization of acrylated Pluronic® F-127.[16] On the other hand, a Pluronic® F-127 double-crosslinked network was prepared *via* physically mixing Pluronic® F-127 gels and carboxymethyl chitosan in the presence of glutaraldehyde. At physiological temperature, the glutaraldehyde crosslinks the carboxymethyl chitosan, causing it to interpenetrate the Pluronic® F-127 gel.[17] Moreno *et al.* used Pluronic® F-127 to conjugate poly(methyl vinyl ether-*co*-maleic anhydride) (Gantrez®) *via* ring opening polymerization. The mechanical and bio-adhesive properties were significantly improved.[19]

10.2.2 Thermogelling Polymers Made from Ring Opening Polymerizations

PEG–oligo-peptides block copolymers are also known as poly(phosphazene) s. This is a class of thermosensitive polymers which are prepared by ring opening polymerization of *N*-carboxy anhydride using methoxy-PEG (mPEG) as initiator. Transition temperatures between 25 °C and 98.5 °C can be obtained by varying the molecular weight of mPEG, molar ratio of the hydrophobic–hydrophilic segments and type of oligo-peptides. These copolymers are enzymatically biodegradable upon injection *in vivo* but are stable during storage in aqueous conditions. For example, PEG-poly(alanine-*co*-phenyl alanine) (PAF) showed thermogelling properties at low concentrations of 3–7 wt% in water.[35] The gelation mechanism is mainly driven by the dehydration of PEG at elevated temperature. Thus, micelles aggregate with hydrophobic peptides to form the core. The transition temperature of this thermogel becomes higher for increasing PEG chains, while lower transition temperatures can be achieved by using more hydrophobic oligo-peptides. Various applications in therapeutic conveyance[35] and wound healing[36] have been reported.

10.2.3 Thermogelling Polyurethanes

Formation of polyurethane is one of the easier ways to prepare thermogelling copolymers. One-pot synthesis is carried out in the presence of polymer diols of low molecular weight and a coupling agent diisocyanate. In the PEG/PPG polyurethane multiblock copolymer, PEG segments are hydrophilic while PPG segments are hydrophilic below 10 °C and hydrophobic above 30 °C. PPG has a LCST of 10–30 °C in water whereby the LCST becomes lower as the molecular weight of the polymer becomes higher.[37] Sol–gel transition in a polyurethane aqueous solution is achieved by self-assembly of hydrophilic and hydrophobic segments into a micellar structure.

Besides PEG and PPG, incorporation of a small amount (1–5%) of a third component – a hydrophobic diol – into the multiblock polyurethane can tune the properties of thermogelling copolymers. For example, PHB was added to lower the CGC of the aqueous system, PCL,[38] poly(tetrahydrofuran carbonate)diol[39] or PLA[40] was added to allow biodegradability, while poly(ethylene butylene)[41] was added to provide bio-stability. The mechanism of sol–gel transition of poly(PEG/PPG/PHB urethane)s was reported in ref. 42, as shown in Figure 10.3. At high dilution (99.9% water, 0.1% polymer), associated micelles are formed by the amphiphilic multiblock copolymers. These self-assembled micelles have PEG hydrophilic tails that interact with water and hydrophobic cores that consist of PPG and PHB. The long polymers chains of PPG and PHB are connected by urethane linkages. A minimum polymer concentration (2–5%), and micelle concentration, is necessary for gel formation. In a testing temperature range of 4–80 °C, the aqueous solution undergoes reversible sol–gel transition. At low temperatures, a clear aqueous solution is formed because the thermosensitive PPG segments are more hydrophilic and the polymers are well-solvated in water. An increase in temperature resulted in the dehydration of PEG segments, and PPG segments become less water-soluble. When a balance of hydrophobicity and hydrophilicity is achieved, a gel state is reached. The driving force for the gel formation is micellar aggregation due to self-association of PEG corona and increased hydrophobicity of PPG. A further increase in temperature resulted

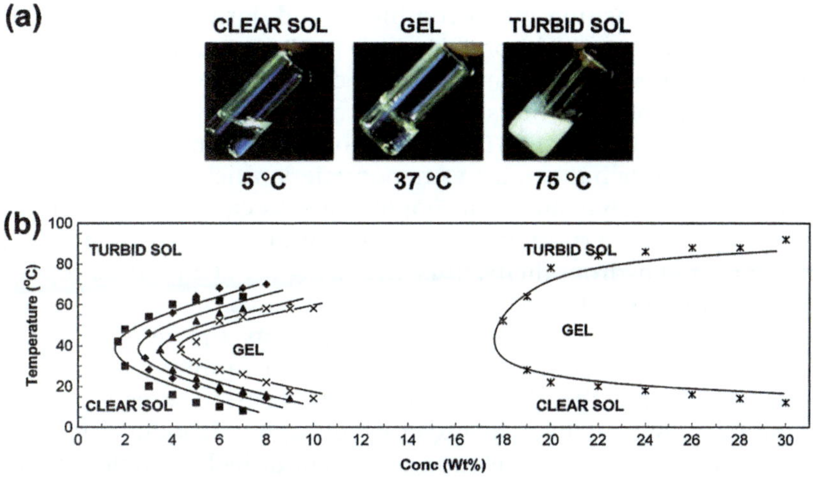

Figure 10.3 (a) Sol–gel transition of poly(PEG/PPG/PHB urethane)s. (b) Phase-diagram of the multiblock polyurethane (left) as compared to Pluronic® F-127 (right). Reproduced with permission from X. J. Loh, S. H. Goh and J. Li, *Biomacromolecules*, 2007, **8**, 585–593. Copyright (2007) American Chemical Society.[43]

in severe dehydration and collapse of the PEG corona, exposing the hydrophobic core and causing a turbid solution. When PHB is added as a third diol in this system, the hydrophobicity is enhanced, leading to a lower CGC as compared to poly(PEG/PPG urethane).

10.2.4 Temperature Responsive PNIPAAm-Based Block Copolymers

PNIPAAm is a pioneer reversible thermosensitive molecule in the development of thermogels. Since the 1960s, numerous studies about the synthesis of thermoresponsive PNIPAAm and its derivatives have been reported for biomedical applications such as therapeutic delivery, cell encapsulation and cell culture sheets.[44] PNIPAAm is well-known for having a low LCST at 32 °C in aqueous solution. More importantly, its LCST is relatively insensitive to changes of pH, concentration or chemical environment.[45] Below its LCST, the polymer is hydrophilic and water-soluble; at the LCST, it exhibits reversible phase change to a hydrophobic state. The polymer structure collapses from coil to globule and precipitates from water.

The ATRP technique is the most effective and widely used method to polymerize (meth)acrylates, (meth)acrylamides, styrene and their copolymers.[46] Hence, ATRP is ideal for the synthesis of most PNIPAAm-based hydrogels. A complex polymer structure (especially graft copolymers) with a narrow polydispersity index can be obtained. The polymerization mechanism involves dynamic equilibrium between the active species activated by redox active transition metal complexes and dormant species. Thermoresponsive PNIPAAm–[hydrophobic core]–PNIPAAm and PNIPAAm–[hydrophilic core]–PNIPAAm copolymers can be obtained by ATRP.[47–50] Compared with PNIPAAm homopolymers, PNIPAAm copolymerized with hydrophobic segments leads to a lower LCST. These copolymers are not suitable for *in situ* gelling applications but can act as nanocarriers which release hydrophobic content at the targeted site as the copolymers precipitate or shrink. On the other hand, copolymerization of PNIPAAm with hydrophilic segments leads to higher overall hydrophilicity, thus increasing the LCST. These copolymers are suitable for injectable *in situ* gelling applications.

Lignin-g-PNIPAAm copolymers, with lignin as the hydrophobic core, precipitate above the LCST (32 °C) rather than forming a physical gel.[48] On the other hand, PNIPAAm–PMPC–PNIPAAm triblock copolymers with hydrophilic poly(2-methacryloyloxyethyl phosphorylcholine) (PMPC) segments form a gel at 37 °C. Gelation is a result of hydrophobic interactions between PNIPAAm blocks at temperatures above the LCST.[50] Hence, the triblock, rather than diblock, conformation is essential for the occurrence of inter-micellar bridging. Recently, Li and coworkers prepared lignin-*b*-PNIPAAm-*b*-(PEG-*co*-PPG) copolymers consisting of lignin as the hydrophobic core, PEG as the hydrophilic corona and PPG and PNIPAAm as thermosensitive segments. These thermosensitive segments transform from hydrophilic to hydrophobic as the temperature becomes higher.[51] The thermogel prepared by ATRP showed sol–gel transition at 33–35 °C and precipitation at 52 °C. Interestingly, the thermogel has a low CGC at 1.3 wt% of polymer in 98.7 wt% of water. Low polymer concentration is advantageous for *in situ* gelling compatibility and cost effectiveness.

RAFT polymerization is another method employed to obtain multiblock amphiphilic copolymers.[52] Compared to multi-step alternating addition of two types of monomers into a living polymerization system, RAFT polymerization is relatively easier to prepare. Narrow polydispersity can be achieved. Using cyclic- or polytrithiocarbonates as the chain transfer agent, a multiblock PNIPAAm–PDMA copolymer was prepared with a two-step addition, as shown in Figure 10.4. The chain length of each sequence can be tuned by the ratio of monomer and trithiocarbonates.[53] A recent study highlighted the feasibility of preparing low LCST blocks of PNIPAAm and PDEAM in an aqueous environment at 25 °C.[54] Moreover, compared to ATRP, RAFT polymerization requires no metal–ligand complex as catalyst for the polymerization; a complex purification procedure can thus be avoided. Double hydrophilic block copolymers (DHBC) have been synthesized using PNIPAAm and poly(*N*,*N*-dimethylacrylamide) (PDMA) *via* a consecutive RAFT polymerization technique.[53,55] The copolymers show thermally-induced unimolecular or multimolecular micelle aggregation, based on their different copolymer architecture, as shown in Figure 10.5. Physical gels are formed when the multiblock PNIPAAm–PDMA consists of PNIPAAm and PDMA with a certain sequence length, due to the formation and aggregation of unimolecular micelles. On the other hand, no gel formation is observed when multimolecular micelles aggregate.

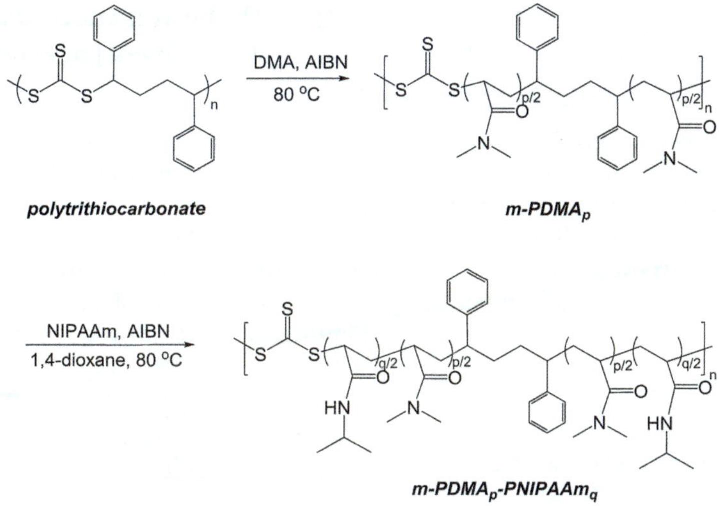

Figure 10.4 Synthesis of multiblock copolymers *m*-PDMA–PNIPAAm by successive RAFT polymerization, using polytrithiocarbonate as chain transfer agent. Reproduced with permission from Z. Ge, Y. Zhou, Z. Tong and S. Liu, *Langmuir*, 2011, **27**, 1143–1151. Copyright (2011) American Chemical Society.[55]

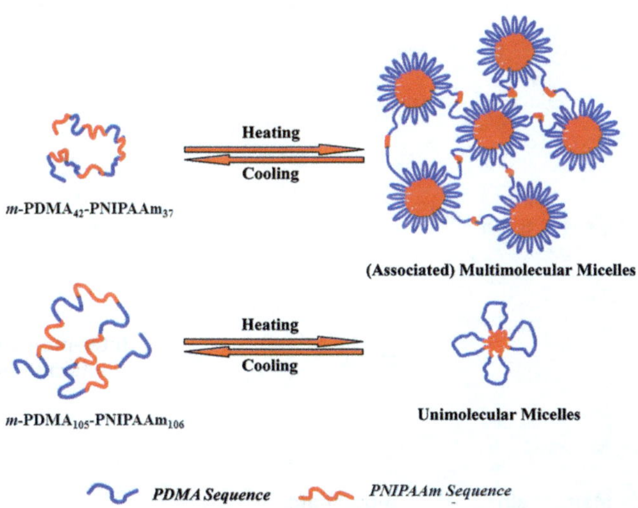

Figure 10.5 Schematic drawing of multiblock PDMA–PNIPAAm showing thermally-induced unimolecular or multimolecular micelle aggregation (red = PNIPAAm, blue = PDMA). Reproduced with permission from Y. Zhou, K. Jiang, Q. Song and S. Liu, *Langmuir*, 2007, **23**, 13076–13084. Copyright (2007) American Chemical Society.[53]

10.2.5 Poly(oligo(ethylene glycol) methyl ether methacrylate) (PoEGMA) and Poly(oligo(ethylene glycol) acrylate Thermogelling Polymers

Lutz *et al.* compared the thermoresponsive properties of PoEGMA copolymers with PNIPAAm, the "gold standard,"[56] and suggested that PoEGMA copolymers can outshine PNIPAAm because of their easily-tunable LCST and biocompatibility comparable to linear PEG. PoEGMA is a relatively new thermoresponsive molecule discovered in the early 2000s. In addition to its thermosensitivity, PoEGMA shows protein-repellant properties which are of great interest as non-fouling surfaces.[57] As shown in Figure 10.6, PoEGMA is a comb-shape

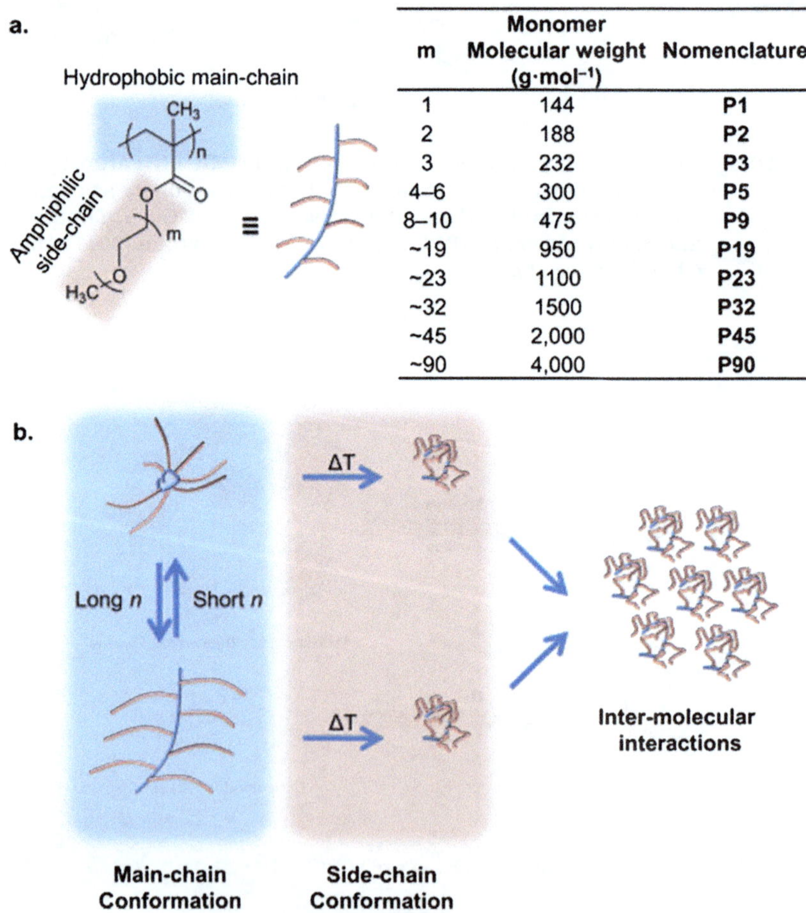

m	Monomer Molecular weight ($g \cdot mol^{-1}$)	Nomenclature
1	144	**P1**
2	188	**P2**
3	232	**P3**
4–6	300	**P5**
8–10	475	**P9**
~19	950	**P19**
~23	1100	**P23**
~32	1500	**P32**
~45	2,000	**P45**
~90	4,000	**P90**

Figure 10.6 (a) Chemical structure of PoEGMA (main and side chains) with different number of repeating units, nomenclature and their molecular weights. (b) Chain length effect: main chain and side chain conformation, and molecular self-assembly at elevated temperature. Reproduced from Liu *et al.*, Conformation-function relationships for the comb-shaped polymer pOEGMA, *Prog. Polym. Sci.*, 2015, **48**, 111–121. Copyright (2015) with permission from Elsevier.[58]

polymer with a hydrophobic backbone and hydrophilic side chains. The LCST is tunable *via* the relative chain length of the main and side chains, as well as the end group functionalities. A few comprehensive reviews recently discussed how the molecular structure can influence the thermoresponsive properties of PoEGMA and PoEGA as biomaterials.[58,59] Well-defined molecular architecture such as polymer brush and amphiphilic block copolymers can be synthesized by controlled radical polymerization including ATRP, NMP (nitroxide-mediated radical polymerization) and RAFT techniques.[60] PoEGMA showed a range of LCST (P2 to P9, in Figure 10.6) from 26–90 °C depending on the number of repeating units at the PEG side chain. The LCST of the copolymer (P2-stat-P9) is tunable *via* the co-monomer composition.[61]

10.3 Evaluating the Resorbability of Thermogels

The biodegradability of thermosensitive hydrogels has received much research attention because it is an essential property for biomedical applications. Introducing biodegradable linkages into the polymer backbone would facilitate the degradation of the copolymer into smaller fragments and its subsequent removal from the body. The biodegradability also affects the efficiency of the treatment because encapsulated therapeutic release is based upon gel degradation and diffusion mechanisms.

The most popular group of thermogels, PEG–PPG–PEG triblock copolymers (or Pluronics®), has been approved by FDA for therapeutic conveyance systems.[62] However, the limitation of PEG-based acrylates and methacrylates is that their carbon–carbon backbone is non-biodegradable. To provide biodegradability, labile linkages or a degradable block can be employed. Biodegradable polymer segments include poly(L-lactic acid) (PLLA), PCL, PHB, PLGA, poly(organophosphazene), poly(propylene fumarate) (PPF) and poly(propylene phosphate).[10,20,22,63–70] Azide- or alkyne-functionalized degradable moieties can also be used. Alkyne-functionalized, star-shaped PEG and trypsin-sensitive degradable polypeptides with azide end groups can form degradable hydrogels.[71]

Recently developed polypeptide-based thermogelling systems demonstrated good *in vitro* stability in aqueous solution and *in vivo* degradability by proteolytic enzymes.[72–74] Genepin has been reported as a crosslinking agent for polymers with amino groups. Genepin crosslinked chitosan was observed to control degradation profile and gel modulus.[75]

Several more approaches can be employed to make biodegradable gels. Hydrogels incorporated with a hyaluronic acid (HA) backbone can be significantly degraded by hyaluronidase. Disulfide-connected multiblock Pluronic® demonstrated a thiol-sensitive degradation and therapeutic release.[76,77] Polymers using acetal as linkers underwent an acid-sensitive degradation and therapeutic release.[78] Thermogelling multiblock Pluronic® modified with MMP-sensitive degradable polypeptide (GPVGLIGK) showed an enzyme-sensitive degradation and therapeutic release.[79] Jeong *et al.* found that the degradation of PEG–PLGA–PEG (550–2810–550) amphiphilic copolymer was affected by the percentage of water and the length of the hydrophobic component.[80] The gels containing more hydrophobic content resulted in slower degradation. Later, Jeong *et al.* reported PEG–PLGA–PEG triblock copolymer that formed transparent gels *in situ* in rats upon subcutaneous injection of 33 wt% aqueous solution.[81] The gel underwent degradation *via* hydrolysis (30% mass loss) and the number average molecular weight decreased to 1900 from 3300 Da after more than a month.

Shim *et al.* added sulfamethazine oligomers (SMOs) to either end of a thermosensitive poly(ε-caprolactone-*co*-lactide)–poly(ethylene glycol)–poly(ε-caprolactone-*co*-lactide) (PCLA–PEG–PCLA) block copolymer to enable pH- and thermo- sensitivity.[82] The block copolymer showed substantially prolonged degradation compared to that of PCLA–PEG–PCLA, due to the buffering effect of sulfonamide groups. Subcutaneous injection of the pH- and thermo- sensitive block copolymer solution (20 wt% in PBS at pH 8.0) into Sprague–Dawley (SD) rats resulted in rapid, stable gel formation, with the injected hydrogel being completely degraded *in vivo* in just six weeks.

Loh *et al.* did extensive research on biodegradable thermogelling polymers for therapeutic release and other biomedical applications. Biodegradable hard segments such as PHB have been introduced into the polyurethane backbone to form poly(ester urethane)s to render the polyurethanes more hydrolytically degradable. These poly(ester urethane)s have been shown to degrade within two weeks under accelerated hydrolytic degradation conditions.[65] Loh *et al.* demonstrated, for the first time, the reversible cyclical thermoresponsive behavior of poly(ester urethane)s comprising of PEG, PPG and PCL.[83] PCL degrades into a naturally occurring metabolite, 6-hydroxyhexanoic acid, by bulk erosion. Loh *et al.* also incorporated poly(lactic acid) (PLA) with PEG and PPG to render the polymer biodegradable.[66] They found that the thermogels hydrolytically degraded to polymer fragments corresponding to the constituent segment blocks within three months. Loh *et al.* performed a detailed study on hydrolytic degradation of multiblock poly(PEG/PPG/PHB urethane) hydrogels in phosphate buffer at pH 7.4 and 37 °C for a period of six months.[84] Using scanning electron microscopy, they found that the hydrogels exhibited increasing porosity with time of hydrolysis (Figure 10.7). The degradation was attributed to random chain scission of the ester backbone bonds of the PHB segments into 3-hydroxybutyric acid monomer and 1–5 units of oligomers. They revealed the mechanism of hydrolytic degradation using GPC, ^{1}H NMR, MALDI-TOF and TGA. The thermogelling copolymer degraded by initial incubation period, followed by the erosion of the polymer gel and the random scission *via* the ester bonds of the PHB segments in buffer solution. The degradation mechanism is similar to that reported for PCLA–PEG–PCLA by Shim *et al.*[82] The rate of chain scission could be controlled by adjusting the copolymer composition.

Loh *et al.*[66] studied hydrolytic degradation of thermogelling poly(PEG/PPG/PLA urethane)s polymers, with PLA content ranging from 1.3 to 6.8 wt%. Aqueous solutions comprising 10 wt% copolymer were injected into a porous cellulose cassette and placed in 25 mL of phosphate buffer release solution at pH 7.4. The degradation process was monitored by analyzing the molecular weight of remaining materials at predetermined time intervals over a period of three months. GPC profiles (Figure 10.8) showed the thermogels degrading into fragments with molecular weights of about 2000, 4000, 6000 and

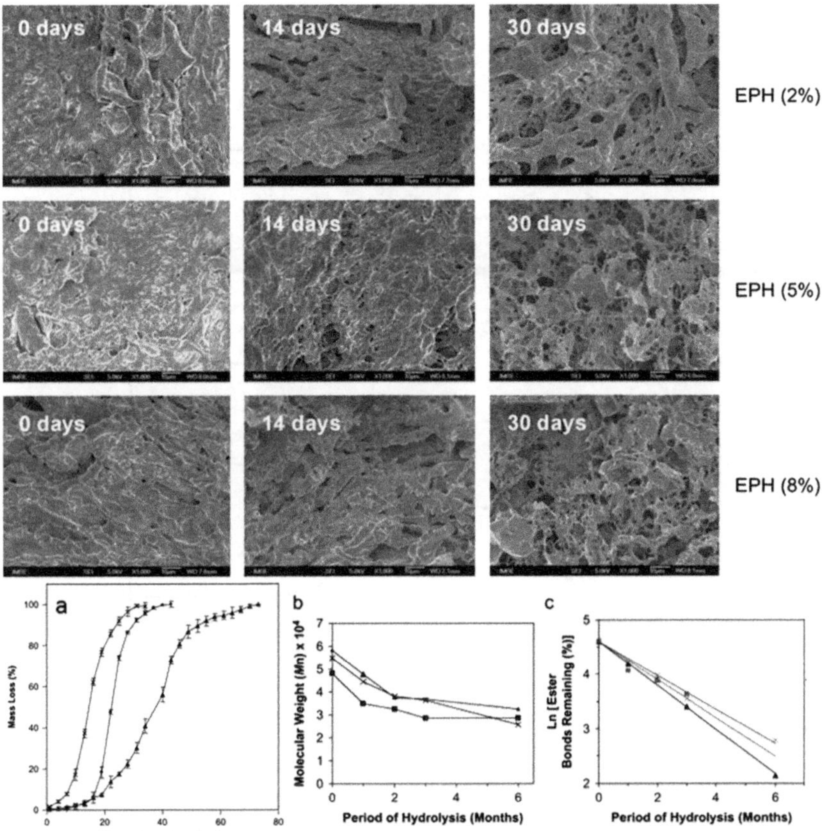

Figure 10.7 SEM images of hydrogel residues after various periods of degradation in PBS at pH 7.4 and 37 °C. Bottom from left to right: (left) mass loss (%) of the poly(PEG/PPG/PHB urethane) hydrogels (5 wt%) after incubation in PBS at pH 7.4 and 37 °C. (middle) Changes in molecular weight of the copolymer degradation products with time of hydrolysis up to 6 months. (right) Plot of the natural logarithm of the fractional ester bonds remaining *versus* degradation time of the polymers after various periods of degradation (▲: EPH (2%), ■: EPH (5%), ×: EPH (8%)). Reproduced from Loh *et al.*, Hydrolytic degradation and protein release studies of thermogelling polyurethane copolymers consisting of poly[(*R*)-3-hydroylbutyrate], poly(ethylene glycol), and poly(propylene glycol), *Biomaterials*, 2007, **28**, 4116–4118. Copyright (2007) with permission from Elsevier.[84]

8000 Da within three months. These fragments can potentially be excreted from the body *via* renal filtration.

Shen and coworkers reported a novel degradable and thermoresponsive PEG analogue polymerized *via* condensation polymerization of PEG-di(meth) acrylates with dithiol.[85] The LCST range is 10–50 °C. The cloud points of these polymers could be tuned by adjusting the PEG chain length and the type of dithiol. DPEG was stable at neutral pH but degraded under acidic conditions as confirmed by hydrolysis experiments (Figure 10.9).

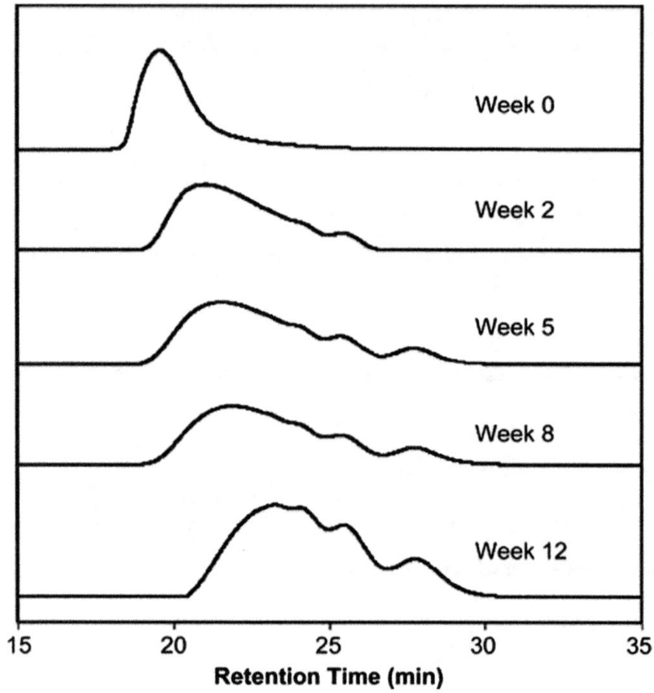

Figure 10.8 GPC profile of remaining poly(PEG/PPG/PLA urethane)s gels incubated in a porous cellulose cassette at various degradation intervals at pH 7.4. Reproduced from Loh *et al.*, Biodegradable thermogelling poly(ester urethane)s consisting of poly(lactic acid) – Thermodynamics of micellization and hydrolytic degradation, *Biomaterials*, 2008, **29**, 2171. Copyright (2008) with permission from Elsevier.[66]

Chu and coworkers developed five types of thermosensitive diblock mPEG–polyester copolymers by ring opening polymerization of methoxy-poly(ethylene glycol) (mPEG) with D,L-lactide, D,L-lactide-glycolide, β-propiolactone, δ-valerolactone and ε-caprolactone respectively.[86] They studied the *in vitro* degradation of mPEG–polyester gels based on weight loss in 30 days at 37 °C and found that the hydrolysis rate of hydrophobic segments primarily determined degradation rate. Among them, mPEG–PLGA degraded the fastest, while mPEG–PPLA showed only 5% weight loss due to its high viscosity. The degradation rate of mPEG-PLGA, mPEG-PVLA and mPEG-PCLA can be found in Figure 10.10. Other types of polymer were not degradable.

Recently, Hsieh *et al.* reported prolonging the *in vitro* and *in vivo* application period of PEG–PLGA–PEG by mixing hydroxyapatite (HA).[87] In this work, they studied the *in vitro* degradation of 15 wt% hydrogels in PBS. The degradation was monitored by measuring the weight of the remaining solid contents for a period of seventy days. The addition of HA prolonged the

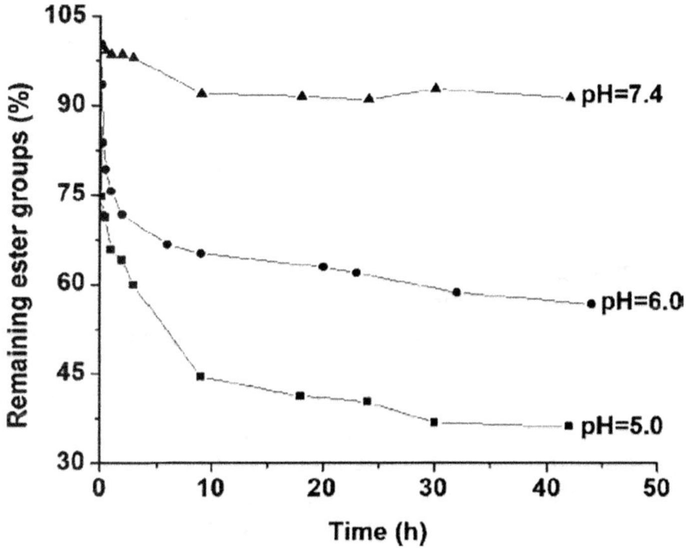

Figure 10.9 *In vitro* hydrolysis of PEGDA258-DET (M_n = 34 100) at 37 °C. Reproduced from Shen *et al.* with permission from John Wiley and Sons © 2007 Wiley Periodical, Inc.[85]

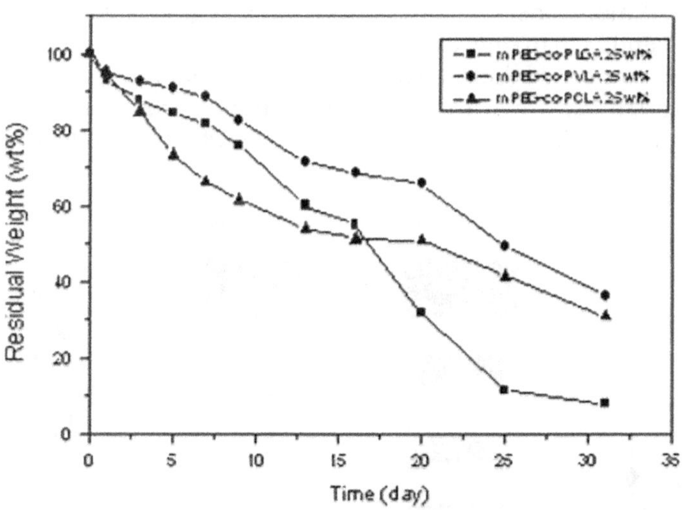

Figure 10.10 Degradation behavior of mPEG–polyester diblock copolymers determined using the weight loss method. Reproduced from Chu *et al.* with permission from John Wiley and Sons © 2010 Society of Chemical Industry.[86]

degradation period and neutralized the solution by increasing the pH. After adding HA, the degradation of the PELGE/HA hydrogel composite was slower than the PELGE hydrogel alone. There was 75 wt% of the hydrogel composite remaining after 70 days. On the other hand, degradation of the PELGE hydrogel was faster in the first 20 days. The biodegradation was studied in mice using dorsal subcutaneous implant PELGE/HA hydrogels and PELGE/HA/BMP-2 hydrogel composites. An inflammatory response appeared at four weeks post-implantation, attributed to the acidic residual monomer PLGA, which has been degraded from the hydrogel. After eight weeks, the absence of inflammatory cells and increase in fibroblasts and new blood vessels indicated release of BMP-2 growth factor from the hydrogel by degradation of the hydrogels.

To date, various biodegradable thermosensitive polymers have been developed. The largest category is modified PEG–PPG copolymers. Other injectable sol–gel transition hydrogels like PLAGA or PLLA/PEG copolymers, PCL/PEG copolymers, PHB/PPG copolymers, PPF/PEG copolymers, polyorganophosphazenes and polypeptides have been investigated.

There are two main types of degradation in polymers – surface erosion and bulk erosion. Surface erosion of polymers proceeds at constant velocity all the time.[88] Bulk erosion of polymers changes erosion velocity with time.[89] The differences are shown in Figure 10.11. Most degradable polymers undergo both surface erosion and bulk erosion. Which of the two processes will dominate is dependent on three factors: (1) the diffusivity of water inside the matrix, (2) the degradation rate of the polymer functional groups and (3) the matrix dimensions.[90,91] For example, PHB degrades *via* surface erosion; PLA, PLGA and PCL mainly exhibit bulk erosion.

The degradation of thermogelling polymers also largely depends on the type of crosslinks, which causes the formation of different chemical entities post-degradation. The properties of degradation products may affect

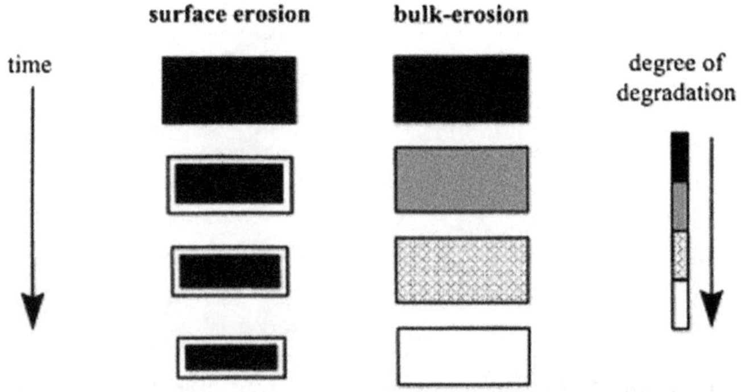

Figure 10.11 Schematic illustration of the changes a polymer matrix undergoes during surface erosion and bulk erosion. Reproduced from Burkersrod *et al.*, Why degradable polymers undergo surface erosion or bulk erosion, *Biomaterials*, 2002, 23, 4221–4231. Copyright (2002) with permission from Elsevier.[91]

the environment after their release. This degradation debris might undergo phagocytosis or pinocytosis. Therefore, the extravasation process should be carefully monitored.

10.4 Therapeutics Encapsulation and Delivery

A thermogelling polymer is responsive to temperature changes resulting in sol–gel transition at body temperature. Thermogels are injectable solutions at lower temperatures, and they turn into harder gel upon injection into a physiological environment. Injectable, biodegradable, *in situ* setting semi-solid thermogels as therapeutic depots and conveyance systems require no follow-up surgical removal after depletion. A hydrophobic therapeutic can be incorporated using the solvent-induced phase inversion technique (SPI). A water-insoluble polymer is first dissolved in an organic, water-miscible solvent containing the therapeutic. Upon injection into the body and exposure to an aqueous environment, the organic solvent dissipates out while water ingresses *via* diffusion.[92] This solvent exchange causes the sol-to-gel transformation and polymer precipitation, leading to implant formation (Figure 10.12).[93]

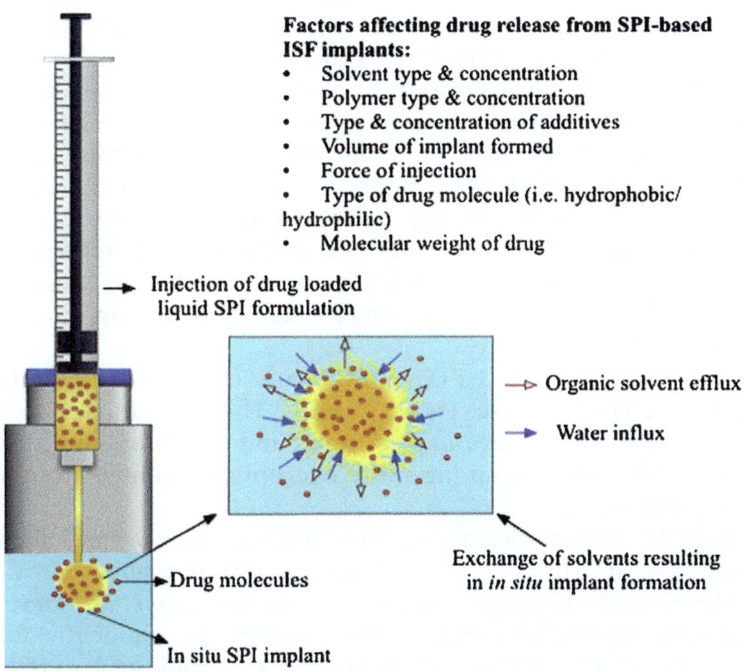

Figure 10.12 Schematic representations of solvent-induced phase inversion technique (SPI) implant formation, solvent exchange and therapeutic delivery. Reproduced from Thakur *et al.*, Solvent induced phase inversion-based *in situ* forming controlled release drug delivery implants, *J. Controlled Release*, 2014, **6**, 8–23. Copyright (2014) with permission from Elsevier.[93]

Thermosensitive PEG–PPG–PEG copolymers (Pluronics®) form a major thermogel family showing an LCST above normal body temperature. They have been extensively applied for therapeutic delivery. For example, Zheng and coworkers[94] developed a Pluronic® analog-based thermosetting gel for ophthalmic therapeutic delivery. Pluronic® analogs were incorporated with mucoadhesive polysaccharide, sodium hyaluronate (HA-Na) for ocular retention. The inclusion of F68 (10%) to F127 (21%) increased the phase transition temperature by 9 °C. The formulation was a free flowing liquid below 25 °C and converted to a firm gel under physiological conditions. Gamma scintigraphic data demonstrated that the precorneal clearance of the thermosetting gel was significantly delayed as compared to the control solution. This means that therapeutic release can be prolonged using these gels. Recently, Guo *et al.* presented an active targeting therapeutic conveyance system with F127-poly(D,L-lactic acid) copolymer decorated with folate ligands.[95] The micelle system was able to kill cancer cells overexpressing folate receptor under low hyperthermia. The micelles based on this F127-PLA copolymer showed an LCST of 39.2 °C. These micelles could remain stable at 37 °C while rapidly releasing encapsulated anticancer therapeutic under low hyperthermia (40 °C, Figure 10.13). With low hyperthermia triggering, this therapeutic conveyance system could significantly amplify the induction of cancer cell apoptosis due to the rapid therapeutic release. These micelles would thus be more effective at elevated temperatures than at 37 °C.

Pluronic® F-127 has been shown to allow sustained therapeutic release of mitomycin C after intraperitoneal injections. Chemotherapeutic efficacy could be kept high while toxicity is reduced even after administering high doses of mitomycin C in F-127.[96] Pluronic® P-85 has been shown to hypersensitize cancer cells expressing multi-therapeutic resistance-associated proteins (MRP1 and MRP2). When P-85 was employed, there was increased intracellular accumulation of MRP-dependent chemotherapeutics. P85 could inhibit MRP therapeutic efflux systems *via* ATP depletion and inhibition of P-glycoprotein ATPase activity.[97]

Jeong *et al.* developed PEG–PLGA–PEG (550–2810–550) amphiphilic copolymer for therapeutic release (Figure 10.14).[80] They found that the degradation was affected by the percentage of water and length of the hydrophobic component. The gels containing more hydrophobic content resulted in a slower degradation rate.

PNIPAAm is another thermosensitive polymer widely studied for therapeutic release. It exhibits a LCST phase separation at about 32 °C, which can be shifted to body temperature by engineering. Loh *et al.*[98] developed a dynamic supramolecular micellar nanocontainer like 'handcuffs' for therapeutic release. The system uses a macrocycle, cucurbit[8]uril, to tether PNIPAAm and poly(dimethyl aminoethyl methacrylate) (PDMAEMA). Without any stimulus, the release profiles of encapsulated doxorubicin (DOX) for both covalent and supramolecular micelles fit Fickian diffusional profiles[84,99] (Figure 10.15). Compared to covalent micelles, the supramolecular micelles have a faster release rate. The application of various triggers increased the

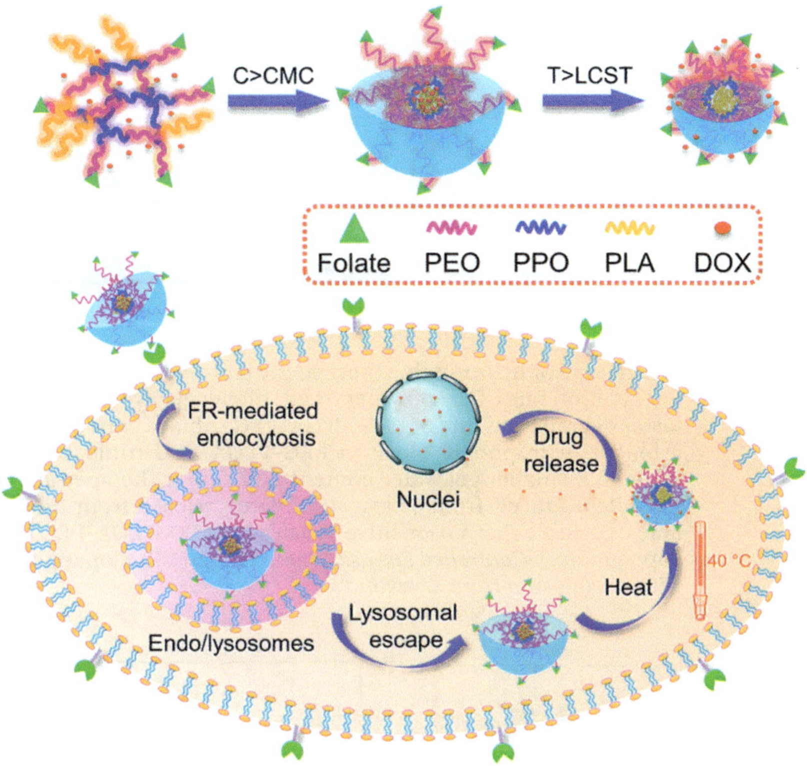

Figure 10.13 Schematic representation of a thermosensitive nanocarrier working as a targeted and controlled therapeutic conveyance system. Reproduced with permission from X. Guo, D. Li, G. Yang, C. Shi, Z. Tang, J. Wang and S. Zhou, *ACS Appl. Mater. Interfaces*, 2014, **6**, 8549–8559. Copyright (2014) American Chemical Society.[95]

release rate, resulting in a controllable reduction of human cervical carcinoma HeLa cell viability.

Recently, Gong and coworkers reported the synthesis of an injectable *in situ*-forming gel named PME, consisting of phospholipids, medium chain triglycerides (MCTs) and ethanol.[100] PME remained in the sol state with low viscosity *in vitro* and turned into a solid or semi-solid gel *in situ* after injection. *In vitro* and *in vivo* doxorubicin release from PME was performed and an initial burst effect was hardly observed from the PME system. Doxorubicin-loaded PME showed anti-proliferative efficacy against MCF-7 breast cancer cells for over 5 days (Figure 10.16). The *in vivo* antitumor activities were evaluated in Kunming mice (male, 22 ± 2 g) with xenograft S180 sarcoma tumors. The sustained release of Dox from PME in tumors was maintained for more than 14 days after one single injection. Hence, this system can be used for localized chemotherapy.

The most popular category of thermogelling polymers is the PEG–PPG–PEG Pluronic® copolymers. Pluronics® are non-ionic copolymer surfactants

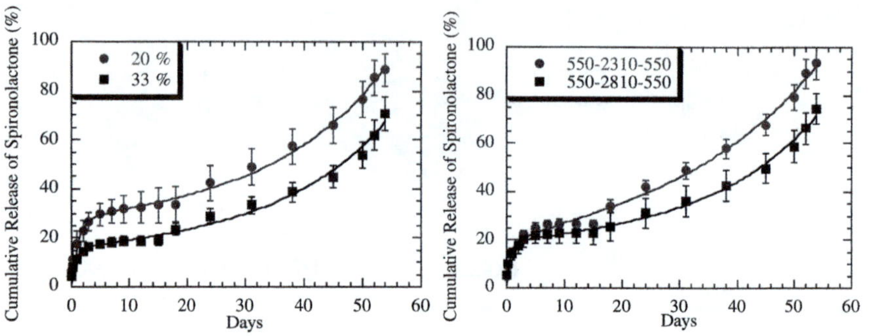

Figure 10.14 The release of hydrophobic therapeutic (spironolactone) reflects the degradation rate of the polymers. Left: higher gel hydration leads to a faster degradation rate of PEG–PLGA–PEG triblock copolymer. Right: a longer hydrophobic block in PEG–PLGA–PEG triblock polymers leads to a smaller gel water content and slower therapeutic release rate. Reproduced from Jeong *et al.*, Drug release from biodegradable injectable thermosensitive hydrogel of PEG–PLGA–PEG triblock copolymers, *J. Controlled Release*, 2000, **63**, 155–163. Copyright (2000) with permission from Elsevier.[80]

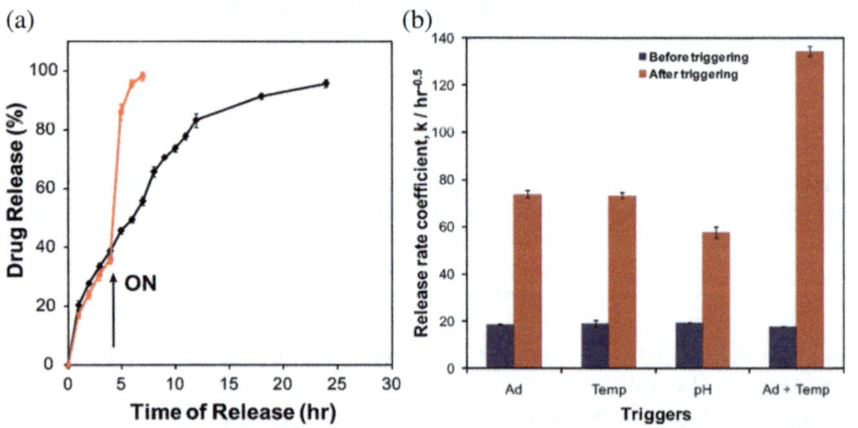

Figure 10.15 (a) Therapeutic release profile from the supramolecular micelles made from 1 + 2 ⊂ CB[8] complex. The red line represents the profile obtained after the system was exposed to the combined adamamtanamine and temperature triggers, while the black line represents the profile obtained without any triggers. (b) Release rate coefficients of the systems after exposure to various triggers. Reproduced with permission from X. J. Loh, J. del Barrio, P. P. C, Toh, T.-C. Lee, D. Jiao, U. Rauwald, E. A. Appel and O. A. Scherman, *Biomacromolecules*, 2011, **13**, 84–91. Copyright (2011) American Chemical Society.[98]

qualified for use in cell culture applications as antifoaming agents. Although they have great potential to be used for therapeutic delivery, some issues should be addressed. Highly concentrated polymer solutions (16 wt%) should be avoided for therapeutic conveyance due to changes in osmolality, transparency and kinetics of gelation of solution. Thermosensitive polymers such

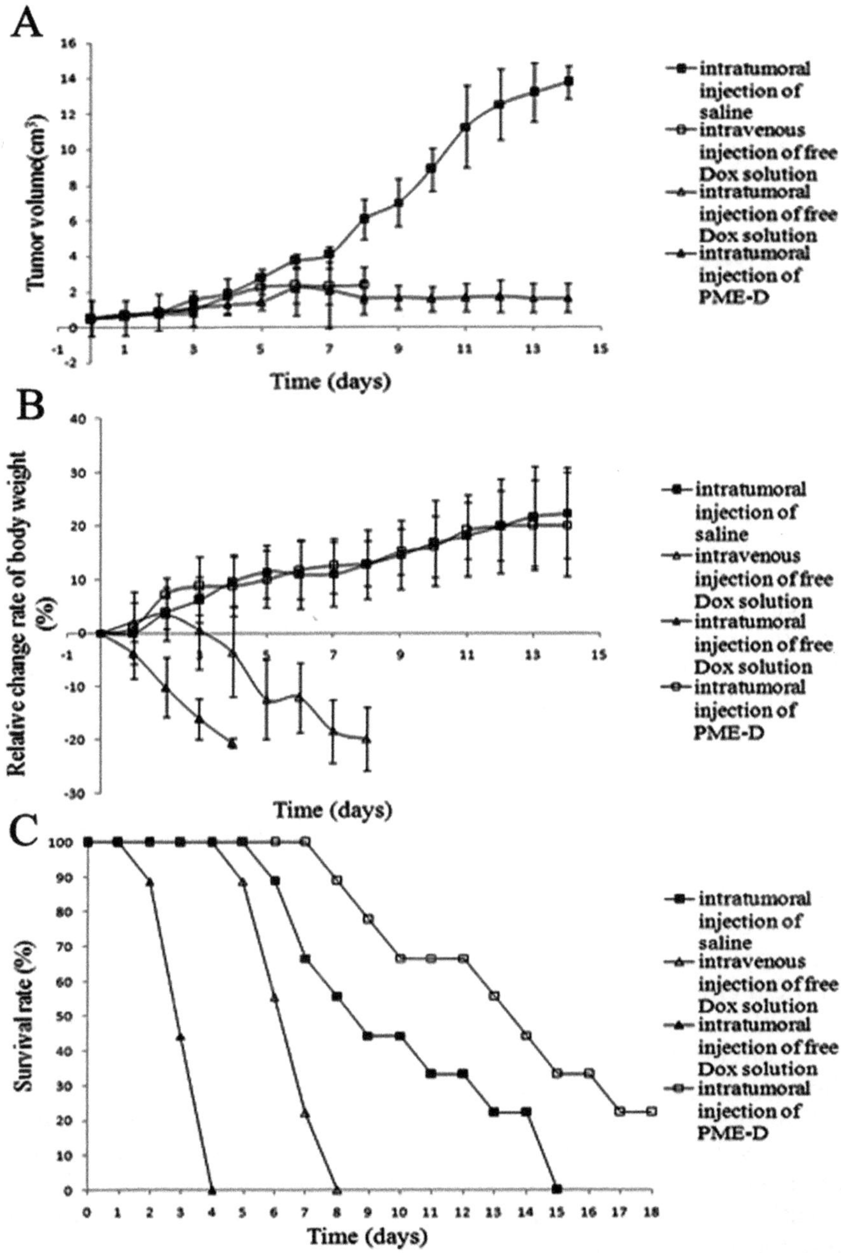

Figure 10.16 *In vivo* antitumor activity in mice bearing S180 sarcoma cancer cell xenografts. The changes of tumor volume (A), relative body weight (B) and survival rate (C) ($p < 0.001$) were monitored to evaluate antitumor activity. Data are represented as the mean ± standard deviation (SD) ($n = 9$). Reproduced with permission from W. Wu, H. Chen, F. Shan, J. Zhou, X. Sun, L. Zhang and T. Gong, *Mol. Pharm.*, 2014, **11**, 3378–3385. Copyright (2014) American Chemical Society.[100]

as PNIPAAm, PVE and PoEGMA also present other challenges. Since a solid gel is not formed immediately upon injection into the body, an initial burst effect of therapeutics can result. Therefore, the polymer should be designed for rapid gel formation. Controlling the molecular weight of the polymer or adding surfactant may be the solution.[101] Both chemical and physical stability of the gel should be considered.

10.5 Outlook and Perspectives

The various creative research referred to in this chapter demonstrates an astounding advancement of thermogel innovation in recent decades. Thermogels can be viably utilized for conveyance of bioactive therapeutics, hereditary material, cells and proteins. This organization technique is exceptionally advantageous and non-obtrusive on the grounds that while the gel is liquid amid blending and infusion, it experiences sol–gel transition at the objective site under physiological conditions. Copolymerization with labile polymers and end group functionalization can give resorbability with the insertion of the correct chemical group.

For remedial conveyance and tissue building, there are stringent necessities for *in situ* thermogelling materials. To start with, the thermogel ought to permit simple detailing and readiness with therapeutics and cells. Second, the material must be non-harmful and biocompatible with the gelling site. Third, the framework ought to permit a tunable and practical restorative discharge profile. The stiffness of thermogels under particular physiological conditions relies on the focused organs, for example, pH, oxidative environment, aggravation, enzymatic impact and protein adsorption. These factors decide the therapeutic conveyance of the gel. Consolidating oxidant moieties, for example, vitamin E, ferulic acid and ascorbic acid, may diminish oxidative stress, while naproxen may help in diminishing aggravation through inflammation. The interpretation of the research to a versatile modern generation is critical. With the thermogelling polymers having LCSTs near body temperature, it is conceivable to sanitize or encourage the polymers with water, as opposed to natural solvents, at temperatures over the LCST. Likewise, an extensive variety of thermogelling polymers stay unexplored (Table 10.1).

Thermoresponsive hydrogels offer reversible sol–gel transition that encourages simple implantation and high adequacy in remedial conveyance of therapeutics. Sooner, rather than later, thermogels can be used for more applications, for example, manufactured vitreous substitutes, eye-drops, wound recuperating patches and skincare items.

References

1. J. Y. Sun, X. Zhao, W. R. Illeperuma, O. Chaudhuri, K. H. Oh, D. J. Mooney, J. J. Vlassak and Z. Suo, *Nature*, 2012, **489**, 133–136.
2. Z. W. Low, P. L. Chee, D. Kai and X. J. Loh, *RSC Adv.*, 2015, **5**, 57678–57685.

3. E. A. Appel, J. del Barrio, X. J. Loh and O. A. Scherman, *Chem. Soc. Rev.*, 2012, **41**, 6195–6214.
4. A. Abdul Karim and X. J. Loh, *Soft Matter*, 2015, **11**, 5425–5434.
5. H. Ye, C. Owh and X. J. Loh, *RSC Adv.*, 2015, **5**, 48720–48728.
6. D. Kai, Z. W. Low, S. S. Liow, A. Abdul Karim, H. Ye, G. Jin, K. Li and X. J. Loh, *ACS Sustainable Chem. Eng.*, 2015, **3**(9), 2160–2169.
7. B. Jeong, S. W. Kim and Y. H. Bae, *Adv. Drug Delivery Rev.*, 2002, **54**, 37–51.
8. M. H. Park, M. K. Joo, B. G. Choi and B. Jeong, *Acc. Chem. Res.*, 2012, **45**, 424–433.
9. H. J. Moon, Y. Ko du, M. H. Park, M. K. Joo and B. Jeong, *Chem. Soc. Rev.*, 2012, **41**, 4860–4883.
10. X. J. Loh and J. Li, *Expert Opin. Ther. Pat.*, 2007, **17**, 965–977.
11. Q. Q. Dou, S. S. Liow, E. Y. Ye, R. Lakshminarayanan and X. J. Loh, *Adv. Healthcare Mater.*, 2014, **3**, 977–988.
12. L. Klouda and A. G. Mikos, *Eur. J. Pharm. Biopharm.*, 2008, **68**, 34–45.
13. C. He, S. W. Kim and D. S. Lee, *J. Controlled Release*, 2008, **127**, 189–207.
14. D. Roy, W. L. Brooks and B. S. Sumerlin, *Chem. Soc. Rev.*, 2013, **42**, 7214–7243.
15. M. A. Ward and T. K. Georgiou, *Polymers*, 2011, **3**, 1215–1242.
16. Y. Shachaf, M. Gonen-Wadmany and D. Seliktar, *Biomaterials*, 2010, **31**, 2836–2847.
17. C. Ju, J. Sun, P. Zi, X. Jin and C. Zhang, *J. Pharm. Sci.*, 2013, **102**, 2707–2717.
18. R. Barreiro-Iglesias, L. Bromberg, M. Temchenko, T. A. Hatton, A. Concheiro and C. Alvarez-Lorenzo, *J. Controlled Release*, 2004, **97**, 537–549.
19. E. Moreno, J. Schwartz, E. Larraneta, P. A. Nguewa, C. Sanmartin, M. Agueros, J. M. Irache and S. Espuelas, *Int. J. Pharm.*, 2014, **459**, 1–9.
20. B. Jeong, Y. H. Bae and S. W. Kim, *Macromolecules*, 1999, **32**, 7064–7069.
21. C. Y. Gong, S. Shi, P. W. Dong, B. Kan, M. L. Guo, X. H. Wang, X. Y. Li, F. Luo, Y. Q. Wei and Z. Y. Qian, *Int. J. Pharm.*, 2009, **365**, 89–99.
22. M. J. Hwang, J. M. Suh, Y. H. Bae, S. W. Kim and B. Jeong, *Biomacromolecules*, 2005, **6**, 885–890.
23. M. Boffito, P. Sirianni, A. M. Di Rienzo and V. Chiono, *J. Biomed. Mater. Res., Part A*, 2015, **103**, 1276–1290.
24. J. Li, X. Li, X. Ni and K. W. Leong, *Macromolecules*, 2003, **36**, 2661–2667.
25. L. Yu, G. Chang, H. Zhang and J. Ding, *J. Polym. Sci., Part A: Polym. Chem.*, 2007, **45**, 1122–1133.
26. L. Yu and J. Ding, *Chem. Soc. Rev.*, 2008, **37**, 1473–1481.
27. G. M. Zentner, R. Rathi, C. Shih, J. C. McRea, M.-H. Seo, H. Oh, B. G. Rhee, J. Mestecky, Z. Moldoveanu, M. Morgan and S. Weitman, *J. Controlled Release*, 2001, **72**, 203–215.
28. L. Yu, H. Zhang and J. Ding, *Angew. Chem.*, 2006, **45**, 2232–2235.
29. G. Chang, L. Yu, Z. Yang and J. Ding, *Polymer*, 2009, **50**, 6111–6120.
30. H. Zhang, L. Yu and J. Ding, *Macromolecules*, 2008, **41**, 6493–6499.

31. L. Chen, T. Ci, T. Li, L. Yu and J. Ding, *Macromolecules*, 2014, **47**, 5895–5903.
32. L. Chen, T. Ci, L. Yu and J. Ding, *Macromolecules*, 2015, **48**, 3662–3671.
33. L. Yu, Z. Zhang, H. Zhang and J. Ding, *Biomacromolecules*, 2010, **11**, 2169–2178.
34. L. Yu, Z. Zhang, H. Zhang and J. Ding, *Biomacromolecules*, 2009, **10**, 1547–1553.
35. Y. Jeong, M. K. Joo, K. H. Bahk, Y. Y. Choi, H. T. Kim, W. K. Kim, H. J. Lee, Y. S. Sohn and B. Jeong, *J. Controlled Release*, 2009, **137**, 25–30.
36. E. J. Yun, B. Yon, M. K. Joo and B. Jeong, *Biomacromolecules*, 2012, **13**, 1106–1111.
37. X. J. Loh, L. W. I. Cheng and J. Li, *Macromol. Symp.*, 2010, **296**, 161–169.
38. X. J. Loh, K. B. Colin Sng and J. Li, *Biomaterials*, 2008, **29**, 3185–3194.
39. X. J. Loh, W. Guerin and S. M. Guillaume, *J. Mater. Chem.*, 2012, **22**, 21249.
40. X. J. Loh, Y. X. Tan, Z. Li, L. S. Teo, S. H. Goh and J. Li, *Biomaterials*, 2008, **29**, 2164–2172.
41. V. P. Nam Nguyen, N. Kuo and X. J. Loh, *Soft Matter*, 2011, **7**, 2150.
42. X. J. Loh, S. H. Goh and J. Li, *J. Phys. Chem. B*, 2009, **113**, 11822–11830.
43. S. H. G. X. J. Loh and J. Li, *Biomacromolecules*, 2007, **8**, 585–593.
44. Y. S. Chen, P. C. Tsou, J. M. Lo, H. C. Tsai, Y. Z. Wang and G. H. Hsiue, *Biomaterials*, 2013, **34**, 7328–7334.
45. H. G. Schild, *Prog. Polym. Sci.*, 1992, **17**, 163–249.
46. K. Matyjaszewski, *Macromolecules*, 2012, **45**, 4015–4039.
47. X. J. Loh, Z.-X. Zhang, Y.-L. Wu, T. S. Lee and J. Li, *Macromolecules*, 2009, **42**, 194–202.
48. Y. S. Kim and J. F. Kadla, *Biomacromolecules*, 2010, **11**, 981–988.
49. X. J. Loh, Y.-L. Wu, W. T. Joseph Seow, M. N. Irzuan Norimzan, Z.-X. Zhang, F.-J. Xu, E.-T. Kang, K.-G. Neoh and J. Li, *Polymer*, 2008, **49**, 5084–5094.
50. C. Li, Y. Tang, S. P. Armes, C. J. Morris, S. F. Rose, A. W. Lloyd and A. L. Lewis, *Biomacromolecules*, 2005, **6**, 994–999.
51. B. Diao, Z. Zhang, J. Zhu and J. Li, *RSC Adv.*, 2014, **4**, 42996–43003.
52. L. Zhang, Q. Wang, P. Lei, X. Wang, C. Wang and L. Cai, *J. Polym. Sci., Part A: Polym. Chem.*, 2007, **45**, 2617–2623.
53. Y. Zhou, K. Jiang, Q. Song and S. Liu, *Langmuir*, 2007, **23**, 13076–13084.
54. L. Martin, G. Gody and S. Perrier, *Polym. Chem.*, 2015, **6**, 4875–4886.
55. Z. Ge, Y. Zhou, Z. Tong and S. Liu, *Langmuir*, 2011, **27**, 1143–1151.
56. J. F. Lutz, O. Akdemir and A. Hoth, *J. Am. Chem. Soc.*, 2006, **128**, 13046–13047.
57. H. Ma, J. Hyun, P. Stiller and A. Chilkoti, *Adv. Mater.*, 2004, **16**, 338–341.
58. M. Liu, J.-C. Leroux and M. A. Gauthier, *Prog. Polym. Sci.*, 2015, **48**, 111–121.
59. G. Vancoillie, D. Frank and R. Hoogenboom, *Prog. Polym. Sci.*, 2014, **39**, 1074–1095.
60. J.-F. Lutz, *J. Polym. Sci., Part A: Polym. Chem.*, 2008, **46**, 3459–3470.

61. J.-F. Lutz and A. Hoth, *Macromolecules*, 2006, **39**, 893–896.
62. http://www.fda.gov/downloads/therapeutics/developmentapprovalprocess/formssubmissionrequirements/therapeuticmasterfilesdmfs/ucm370723.txt.
63. H. F. Naguib, M. S. A. Aziz, S. M. Sherif and G. R. Saad, *Appl. Clay Sci.*, 2012, **57**, 55–63.
64. M. S. Abdel Aziz, H. F. Naguib, S. M. Sherif and G. R. Saad, *Polym.-Plast. Technol. Eng.*, 2014, **53**, 1671–1681.
65. X. J. Loh, K. K. Tan, X. Li and J. Li, *Biomaterials*, 2006, **27**, 1841–1850.
66. X. J. Loh, Y. X. Tan, Z. Li, L. S. Teo, S. H. Goh and J. Li, *Biomaterials*, 2008, **29**, 2164–2172.
67. M. K. Joo, Y. S. Sohn and B. Jeong, *Macromolecules*, 2007, **40**, 5111–5115.
68. B. Jeong, Y. H. Bae, D. S. Lee and S. W. Kim, *Nature*, 1997, **388**, 860–862.
69. S. J. Bae, J. M. Suh, Y. S. Sohn, Y. H. Bae, S. W. Kim and B. Jeong, *Macromolecules*, 2005, **38**, 5260–5265.
70. X. J. Loh, S. H. Goh and J. Li, *Biomacromolecules*, 2007, **8**, 585–593.
71. M. van Dijk, C. F. van Nostrum, W. E. Hennink, D. T. S. Rijkers and R. M. J. Liskamp, *Biomacromolecules*, 2010, **11**, 1608–1614.
72. Y. Jeong, M. K. Joo, K. H. Bahk, Y. Y. Choi, H.-T. Kim, W.-K. Kim, H. J. Lee, Y. S. Sohn and B. Jeong, *J. Controlled Release*, 2009, **137**, 25–30.
73. B. G. Choi, M. H. Park, S.-H. Cho, M. K. Joo, H. J. Oh, E. H. Kim, K. Park, D. K. Han and B. Jeong, *Biomaterials*, 2010, **31**, 9266–9272.
74. H. J. Moon, B. G. Choi, M. H. Park, M. K. Joo and B. Jeong, *Biomacromolecules*, 2011, **12**, 1234–1242.
75. M. J. Moura, H. Faneca, M. P. Lima, M. H. Gil and M. M. Figueiredo, *Biomacromolecules*, 2011, **12**, 3275–3284.
76. D. Y. Ko, U. P. Shinde, B. Yeon and B. Jeong, *Prog. Polym. Sci.*, 2013, **38**, 672–701.
77. K. H. Sun, Y. S. Sohn and B. Jeong, *Biomacromolecules*, 2006, **7**, 2871–2877.
78. V. K. Garripelli, J.-K. Kim, R. Namgung, W. J. Kim, M. A. Repka and S. Jo, *Acta Biomater.*, 2010, **6**, 477–485.
79. V. K. Garripelli, J.-K. Kim, S. Son, W. J. Kim, M. A. Repka and S. Jo, *Acta Biomater.*, 2011, **7**, 1984–1992.
80. B. Jeong, Y. H. Bae and S. W. Kim, *J. Controlled Release*, 2000, **63**, 155–163.
81. B. Jeong, Y. H. Bae and S. W. Kim, *J. Biomed. Mater. Res.*, 2000, **50**, 171–177.
82. W. S. Shim, J.-H. Kim, H. Park, K. Kim, I. Chan Kwon and D. S. Lee, *Biomaterials*, 2006, **27**, 5178–5185.
83. X. J. Loh, K. B. C. Sng and J. Li, *Biomaterials*, 2008, **29**, 3185–3194.
84. X. J. Loh, S. H. Goh and J. Li, *Biomaterials*, 2007, **28**, 4113–4123.
85. N. Wang, A. Dong, M. Radosz and Y. Shen, *J. Biomed. Mater. Res., Part A*, 2008, **84**, 148–157.
86. C. F. Chen, C. T. Y. Lin and I. M. Chu, *Polym. Int.*, 2010, **59**, 1428–1435.
87. L. Mei-Chun, C. Kung-Chin, H. Sheng-Chieh, C. Ming-Chieh, I. H. Wei, H. You-Rong, L. Hung-Maan, H. Ming-Fa and Y. Jui-Ming, *Biomed. Mater.*, 2014, **9**, 015011.

88. A. Gopferich and R. Langer, *Macromolecules*, 1993, **26**, 4105–4112.
89. A. Göpferich, *Macromolecules*, 1997, **30**, 2598–2604.
90. S. N. Rothstein, W. J. Federspiel and S. R. Little, *Biomaterials*, 2009, **30**, 1657–1664.
91. F. v. Burkersroda, L. Schedl and A. Göpferich, *Biomaterials*, 2002, **23**, 4221–4231.
92. P. D. Graham, K. J. Brodbeck and A. J. McHugh, *J. Controlled Release*, 1999, **58**, 233–245.
93. R. R. S. Thakur, H. L. McMillan and D. S. Jones, *J. Controlled Release*, 2014, **176**, 8–23.
94. G. Wei, H. Xu, P. T. Ding, S. M. Li and J. M. Zheng, *J. Controlled Release*, 2002, **83**, 65–74.
95. X. Guo, D. Li, G. Yang, C. Shi, Z. Tang, J. Wang and S. Zhou, *ACS Appl. Mater. Interfaces*, 2014, **6**, 8549–8559.
96. S. Miyazaki, Y. Ohkawa, M. Takada and D. Attwood, *Chem. Pharm. Bull.*, 1992, **40**, 2224–2226.
97. E. Batrakova, S. Li, V. Alakhov, W. Elmquist, D. Miller and A. Kabanov, *Pharm. Res.*, 2003, **20**, 1581–1590.
98. X. J. Loh, J. s. del Barrio, P. P. C. Toh, T.-C. Lee, D. Jiao, U. Rauwald, E. A. Appel and O. A. Scherman, *Biomacromolecules*, 2011, **13**, 84–91.
99. X. J. Loh, V. P. N. Nguyen, N. Kuo and J. Li, *J. Mater. Chem.*, 2011, **21**, 2246–2254.
100. W. Wu, H. Chen, F. Shan, J. Zhou, X. Sun, L. Zhang and T. Gong, *Mol. Pharm.*, 2014, **11**, 3378–3385.
101. M. Madan, A. Bajaj, S. Lewis, N. Udupa and J. A. Baig, *Indian J. Pharm. Sci.*, 2009, **71**, 242–251.

CHAPTER 11

Pectin As a Rheology Modifier: Recent Reports on Its Origin, Structure, Commercial Production and Gelling Mechanism

SIEW YIN CHAN[a,b], WEE SIM CHOO*[a], DAVID JAMES YOUNG*[a,b,c], AND XIAN JUN LOH*[b,d,e]

[a]School of Science, Monash University Malaysia, Bandar Sunway 47500, Selangor, Malaysia; [b]Institute of Materials Research and Engineering (IMRE), A*STAR, 3 Research Link, Singapore 117602, Singapore; [c]Faculty of Science, Health, Education and Engineering, Maroochydore DC, Queensland 4558, Australia; [d]Department of Materials Science and Engineering, National University of Singapore, 9 Engineering Drive 1, Singapore 117576, Singapore; [e]Singapore Eye Research Institute, 11 Third Hospital Avenue, Singapore 168751, Singapore
*E-mail: choo.wee.sim@monash.edu, dyoung1@usc.edu.au, lohxj@imre.a-star.edu.sg

RSC Polymer Chemistry Series No. 20
Polymers for Personal Care Products and Cosmetics
Edited by Xian Jun Loh

Published by the Royal Society of Chemistry, www.rsc.org

11.1 Introduction

Pectin is well-known for its gelling properties. This polysaccharide was first discovered and named by Hennri Braconnot (1825), borrowing from the Greek word 'pektikos,' meaning to congeal or solidify. Braconnot proposed that pectins should have important functions in plants and scientists have pursued the study of this macromolecule vigorously since then. Pectin has been widely used in a variety of applications, including in food, pharmaceutical material, personal care products and also in the polymer industry.[1–6] It is used as a gelling and stabilizing agent in the food and cosmetic industries and it has been shown to have numerous positive influences on the health of humans including lowering cholesterol and serum glucose levels, reducing cancer propensity and stimulating the immune response.[7] It is also used in the production of a variety of specialty products including edible and biodegradable films, adhesives, paper substitutes, foams and plasticizers, surface modifiers for medical devices, materials for biomedical implantation and for drug delivery.[7] These unique functions arise in part from the multiplicity of structural epitopes.[7] What is pectin? The term 'pectin' is somewhat misleading because it implies one molecule; but pectin is a family of oligosaccharides and polysaccharides that have common features, yet are extremely diverse in their fine structures.[8] In this chapter, we use the singular pectin and plural pectins interchangeably.

11.2 Biological Aspects of Pectins

Pectins are natural polysaccharides derived from plants that make up about one third of the cell wall materials of dicotyledonous and some monocotyledonous plants. Pectins can also be found in small proportions in cell walls of grasses. In plant tissue, pectins are often associated with other cell wall components such as cellulose, hemicellulose or lignin.[9] Together, they play important roles in plant growth and development. Pectins help cells to adhere to each other, providing consistency and mechanical resistance to cell tissues, behaving in the manner of stabilized gels. Middle lamella contains the highest concentrations of pectins, with a gradual decrease of pectin concentration as one passes through the primary cell wall towards the plasma membrane.[10]

Pectins play a role in plant growth, development, morphogenesis, defense, cell–cell adhesion, wall structure, signaling, cell expansion, wall porosity, binding of ions, growth factors and enzymes, pollen tube growth, seed hydration, leaf abscission and fruit development.[11] They have different specific functions depending on their locations in the cell walls or plant tissues and the amount and the nature of pectic molecules present. Pectins function as hydrating agents and cementing material for the cellulosic network[7] and can be brought into solution more easily compared to other cell wall polymers. Pectins therefore function as a gel, comprising a network that can trap solute molecules within it.[12]

11.3 Chemical Aspects of Pectin

Although it has been more than 100 years since the discovery of pectin, the chemical and structural properties of native pectin are still a subject of investigation because of the inhomogeneity of this polymer. The chemical structure of pectin varies among different plant species and even within the same species depending on the growing environment.[13] Pectin is a group of complex polysaccharides that contains covalently 1,4-linked α-D-galacturonic (galactosyluronic) acid residues (Figure 11.1). It is thought that the native pectin in plants is largely composed of three polysaccharide domains: homogalacturonan (HGA), rhamnogalacturonan I (RG-I), and rhamnogalacturonan II (RG-II)[14,15] while some studies proposed that the latter is a type of substituted HGAs.[16] These three polysaccharide domains are believed to be covalently linked to form a pectic network throughout the plant tissues.[15,17] However, the way in which the different domains are joined to one another to form the pectin macromolecule remains unclear.[16]

HGA is a linear homopolymer of α-(1 → 4)-D-galacturonic acid (Figure 11.2a). It is estimated to contain around 100–200 GalA units.[18] HGA is an abundant and widespread domain of pectin. It is known as the linear region (Figure 11.3b). The polymer appears to be synthesized in the Golgi apparatus and then transferred to the middle lamella and primary cell walls.[15] Pectin is mostly deposited in the cell wall as a form that has 70–80% of GalA residues that are methylated at the C-6 carboxyl groups (Figure 11.2b).[15] These GalA residues may be O-acetylated as well, depending on the plant; this occurs predominantly at the C-3 position although C-2 substitution can also occur (Figure 11.2c).[15] The linear region can sometimes be joined by one or two α-(1 → 2)-L-rhamnopyranose residues and most of the pectin backbones have this structure.[15] GalA residues may be substituted at C-2 or

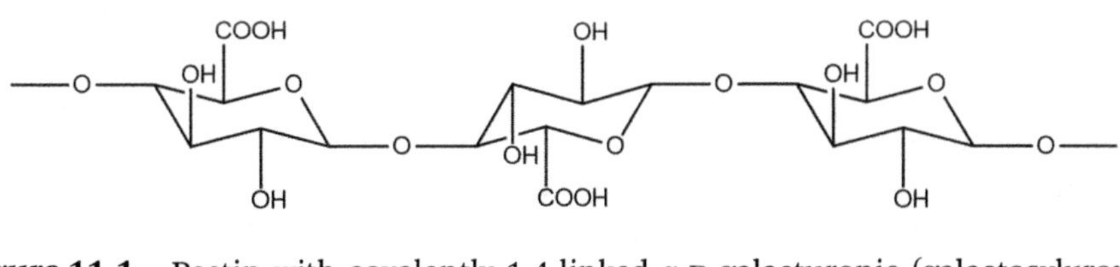

Figure 11.1 Pectin with covalently 1,4-linked α-D-galacturonic (galactosyluronic) acid.

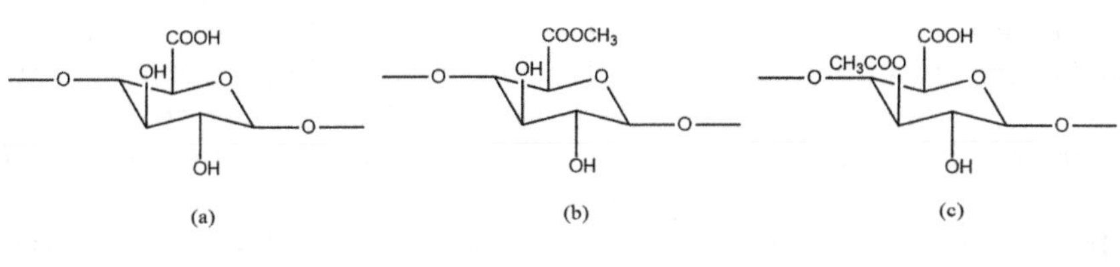

(a) (b) (c)

Figure 11.2 (a) Galacturonic acid. (b) Methylated galacturonic acid. (c) Acetylated galacturonic acid.

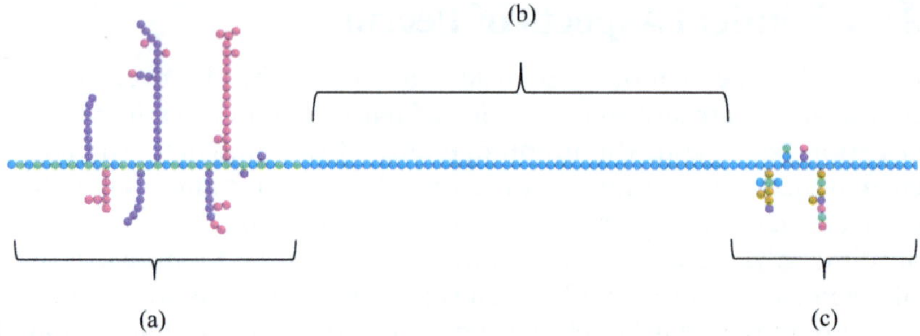

Figure 11.3 The pectin chain comprising of three covalently linked pectin polysaccharides: (a) rhamnogalacturonan I (RG-I), (b) homogalacturonan (HGA), and (c) rhamnogalacturonan II (RG-II).[15]

C-3 with residues of xylose or apiose, producing domains known as xylogalacturonan or apiogalacturonan, respectively.[19] These biosynthetic modifications, or combinations of modifications, are likely to alter the functional properties of the HGA domain.[7,8,15,17] However, substituted HGAs such as xylogalacturonan and apiogalacturonan are less frequently occurring building blocks of pectin.[16]

The RG-I backbone contains up to 100 or more repeating units of the disaccharide α-(1 → 2)-L-rhamnose-α-(1 → 4)-D-galacturonic acid (Figure 11.3a).[19] In most cases, 20–80% of rhamnose residues in RG-I are substituted at C-4 with side chains in which neutral residues predominate, and these can vary in size from a single glycosyl residue to 50 or more, resulting in a large and highly variable family of polysaccharides.[7] Common structural features of the side chains include polymeric β-(1 → 4)-D-galactosyl and α-(1 → 5)-L-arabinosyl residues.[15] A range of other linkages involving other sugars or uronic acids can also be present. Pectic β-(1 → 4)-D-galactan with non-reducing terminal-arabinose (t-Ara) substituted at the O-3 of some of the GalA units is known as type I arabinogalactan.[15] The highly branched nature of RG-I has resulted in it being known as the hairy region. Methyl-esterification of GalA residues within RG-I is not known.

RG-II is not structurally related to RG-I, despite their assigned names, and it is also included in the hairy region. It is a branched pectic domain containing an HGA backbone. It is highly compact and is composed of an HGA backbone with around 9 GalA units that are α-(1 → 4) linked and to which 4 structurally different polymeric side chains of known and consistent lengths are substituted (Figure 11.3c).[15] It is thought that the ends of RG-II are glycosidically linked to the HGA domains.[20] Apart from rhamnose, the side chains contain 11 rare sugars including apiose, 3-O-methyl-L-fucose, 2-O-methyl-D-xylose, 3-C-carboxy-5-desoxy-L-xylose (aceric acid), 3-deoxy-D-manno-octulosonic acid (KDO) and 3-deoxy-D-lyxoheptulosaric acid.[20] An interesting feature of RG-II is that it is involved in cross-linking two pectin molecules (apiosyl residues) within the cell wall by borate ester links.[15,19] It is a highly conserved and widespread domain isolated from cell walls by endopolygalacturonase cleavage, indicating its covalent attachment to HGA.[15,19] However, RG-II appears to be the only major pectic domain that does not have significant structural diversity or modulation of its fine structure.[15]

11.4 Galacturonic Acid Units and Degree of Esterification

Pectin is generally recognized as safe (GRAS) by the United States Food and Drug Administration (FDA). Aside from their gelling, stabilizing and thickening properties, commercial pectins are extensively used as functional food ingredients with possible health promoting effects. Pectin is listed as 440 by Codex Alimentarius and E440 by the European Union (EU) code.[21] The chemical structure of pectin correlates with its functionality and applications.

The Food and Agriculture Organization (FAO) and European Union (EU) have determined that standard commercial 'pectin' must contain at least 65% GalA units. Pectic substances from different sources or individual fruits differ greatly in their properties and behavior, although galacturonic acids are the main constituent of all pectins. These variations in properties are believed, to a large extent, to be related to the variation in galacturonic acid composition. The carboxylic groups and hydroxyls of GalA monomers in their molecular structure may or may not be methylated or acetylated, respectively. The methylated and acetylated groups are expressed as the degree of methylation (DM) and degree of acetylation (DA), respectively. There are other more detailed specifications that can distinguish the wide range of pectin types, including the presence, length, branching of neutral sugar side chains and the molecular masses. However, in the food industry, DM is used as the general classification of pectin and DA is used occasionally. The DM, to a large extent, determines the ion exchange and water-binding capacity of pectin and the possibility of cross-linking with the formation of salt bridges, ester linkage or hydrogen bonding.[15]

There are two types of pectin: high methoxyl (HM) pectin with DM > 50% and low methoxyl (LM) pectin with DM < 50%. HM and LM pectins have different physicochemical characteristics and therefore different applications. HM pectins form gels in acidic systems (pH 2.0–3.5) with the presence of large concentrations (55–75%) of co-solutes such as sucrose or sorbitol, while LM pectin can gel with the addition of divalent ions, such as calcium, in the absence of co-solutes, particularly sugar, over a wide range of pH (pH 2.0–6.0).[22] It is proposed that each pectin molecule is highly esterified when pectin is first synthesized. However, once transferred to the cell wall, pectin methylesterases in the plant de-esterify the GalA units slowly, resulting in a degree of methyl-esterification of the galacturonic acid units between 60–90%, depending on the origin of the pectin, the state of the cell wall and the maturity of the plant.[9,23] Commercial pectin has a typical DM between 55–75%, depending on the plant species and maturity and also the extraction conditions.[24] Commercial extraction and subsequent modification usually give rise to less complex structures because of de-esterification and de-polymerization of pectin molecules.[9]

11.5 Commercial Pectin

It has been reported that pectin contributes significantly to the fast-growing global market for hydrocolloids. About $850 million worth of pectin was sold in 2013 and this figure is expected to continue increasing at 5–6% per year.[25] The world's largest pectin producer by volume is the food ingredients' company, CP Kelco. Commercial pectin is usually extracted from by-products of the food industry such as citrus peel, apple pomace, and to a smaller extent, sugar beet pulp. Citrus peel and apple pomace are available in copious supply as remainders from juices production,[26] while sugar beet pulp is obtained from the sugar industry.[27] Other by-products, including soy hull and passion fruit peels, have been investigated as sources of pectin, but the former apple, citrus and beet have higher pectin content.[23] Apple pectin produces a more viscous gel and it is mainly used as bakery fillings.[27,28] Citrus pectin produces a lighter color of gel and thus it is more likely to be applied in jelly and confectionery productions.[27] In contrast, pectin extracted from sugar beet pulp has poor gelling properties.[27] However, sugar beet pectin was reported to act as an effective emulsifier in oil-in-water emulsions.[23,29]

11.6 Industrial Production of Pectin

The commercial extraction of pectin occurs in three main stages: the acid aqueous extraction, followed by isolation of pectin and finally the characterization of the pectin.[28,30] Acids or, to a smaller extent, other additives such as enzymes are employed to give a higher production yield or better quality pectin compared to alkali or water alone.[10,31,32]

A typical extraction process includes subjecting the raw materials to a pre-treatment process such as blanching, washing or drying. This pre-treatment is carried out to inactivate enzymes that otherwise would rapidly degrade the pectin and to increase the product stability during transportation and storage.[23] The raw materials are then treated with acid (pH 1.5 to 3.0) at high temperature (70 to 90 °C). The commonly used acids are mineral acids such as hydrochloric acid,[9] nitric acid[29,33,34] or sulfuric acid.[35,36] These acids help to solubilize pectic materials from the plant tissues, facilitating the extraction process. However, under the acidic conditions, degradation reactions such as de-esterification and de-polymerization may occur, affecting the properties of pectin. Therefore, extraction conditions such as temperature, time or pH must be carefully controlled.[23,30]

The extract containing pectin will then be separated from the raw materials by filtration or centrifugation.[23,35,37] The pectin will be precipitated from the purified extract using alcohols isopropanol, ethanol or methanol.[9] Insoluble salts such as polyvalent cations, usually aluminium, can be used to precipitate pectins as well.[9] The precipitates obtained will be washed with alcohol again and pressed to remove impurities.

The final step involves drying and milling to yield powdered pectin.[9] The pectins obtained will be blended with different production batches to standardize the variability of pectin in structure and functionality from different natural sources.[24] Dilution of pectin with sucrose or dextrose may be done to attain a standard performance.[24] Most of the pectins obtained in commercial processes are HM pectins, approximately DM 70% or less.[24]

Recently, there have been reports of ultrasound, microwave and electric field assisted methods for the extraction of pectin.[38–40] The pectin extracted by ultrasound possessed lower viscosity, molecular weight and degree of esterification, but with a greater degree of branching and purity than conventional heat-extracted pectin. On the other hand, microwave extraction can provide yields of about 30%. Electric field, on the other hand, is used to shorten the extraction time by heat and to reduce the degradation brought about by the extended exposure to heat using the conventional method.

11.7 The Influence of Extraction Conditions on the Isolation and Recovery of Pectin

The chemical agents used for pectin extraction can be divided into four groups: water and buffers, ion chelators, acids and bases. When used successively in the above-mentioned order (from water to base), these different chemical agents may selectively extract pectic materials from the same starting cell wall material. Each pectin extract has a different chemical make-up due to solubility differences.[41] However, this selectivity is less evident when the extraction does not occur in this sequence, especially when acid is used first or when these agents are used individually to extract pectin from the same type of materials.[42]

Acids are generally the highest yielding extracting agents.[42] The most commonly used organic acids are acetic, citric, lactic, malic and tartaric acid.[42,43] However, on a commercial scale, pectin from either apple pomace or citrus peel is extracted by treating the raw material with hot dilute mineral acid. The extraction parameters substrate–extractant weight ratio, pH, temperature and time are generally in the range of 1 : 15–1 : 35 (w/v), pH 1.4–3.0, 60–100 °C and 20–360 min, respectively.[27,44] Acid extraction of pectin has many advantages, including high extraction yield, and the pectin obtained is generally enriched with the main constituent of galacturonic acid.[9,42] The esterified methoxyl groups at C-6 of galacturonic acid units within the homogalacturonic backbone permit good gelation in the presence of highly soluble solids such as sucrose and acid under specified conditions.[42]

11.8 Production of Low Methoxyl Pectin

Although LM pectin does occur naturally in plants, it is usually obtained commercially from HM pectin. The process of preparing LM pectin from HM pectin is commonly known as de-esterification. LM pectins are usually manufactured from HM pectins by controlled acid de-esterification, alkaline de-esterification, enzyme de-esterification or ammonia de-esterification.[42] Acid, alkali or enzyme de-esterification yield LM pectins with carboxyl or methoxyl groups on C-6, whereas the ammonia system yields LM pectin with acid amide groups in addition to carboxyl and methoxyl groups on C-6.[45,46] Among these methods, de-esterification using ammonia or a combination of ammonia treatments with the others (acid, alkali or enzymatic) are more commonly used commercially.[45] The selection and control of the extraction conditions is important to determine the yield and quality of LM pectin. In terms of yield, there might be concomitant degradation during the de-esterification and this will be covered in detail later. In terms of quality, there are two types of carboxyl distribution that can be produced by different de-esterification methods: random and blockwise.[47] Random is defined as the homogeneous distribution of carboxylic acids in the LM pectin while blockwise refers to heterogeneous distribution of carboxylic acids.

11.8.1 Acid De-Esterification

Acid de-esterification of pectins occurs randomly in the methoxyl groups along the galacturonan backbones.[48] The glycosidic bonds of uronic acid units are reported to be resistant to acid hydrolysis at relatively low pH,[46] although more recent studies indicate that the reaction rate of acid hydrolysis increases with decreasing pH.[48–50] It has been proposed that the observed acid hydrolysis of pectin is actually the removal of neutral sugar side chains, or 'ballast'. Taylor[51] described 'ballast' as 'sugars excluding galacturonic acid' while Panchev *et al.*[52] describe it as 'mainly hemicellulose'. The removal of ballast during de-esterification yields pectins with a higher percentage of GalA units.[52] In acid de-esterification, the hydrolysis of glycosidic linkages is most likely happening in the neutral sugar side chains, which may lead to an increase of the GalA content and a decrease of the neutral sugar content of the treated pectin.[46,51] The increase of the GalA content is important for determining the DM values of pectin and could be advantageous or disadvantageous. On the positive side, it contributes to a higher ratio of GalA units to methylated GalA units, yielding a lower degree of methyl-esterification and greater gelling. On the negative side, the decreased molecular weight of pectin will possibly influence the strength of the gel formed.[9]

Acid de-esterification has a relatively low rate of reaction even though strong acid, particularly hydrochloric acid (HCl), is commonly used.[53] At controlled or reduced temperature, there is a higher possibility that LM pectins will be obtained without extensive main-chain breakdown and a concomitant decrease of the molecular weight of the pectin.[46] Higher temperature

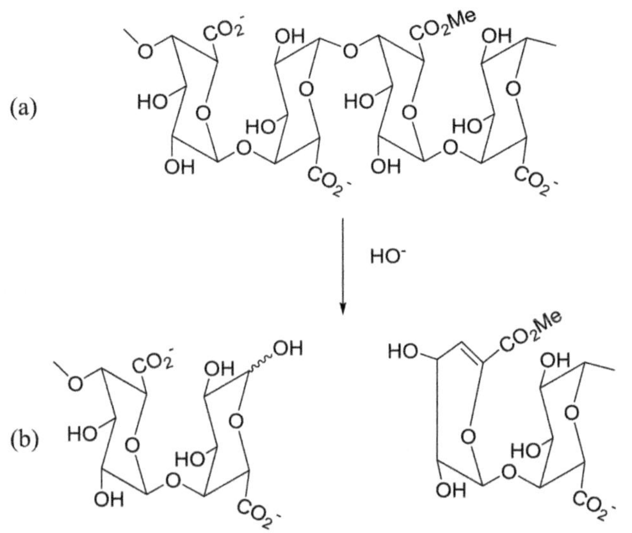

Figure 11.4 β-Elimination mechanism of methylated galacturonic acid. (a) Cleavage of glycosidic bonds in the β-position to the carbonyl group. (b) Formation of a glycosyl anion and a double bond between C-4 and C-5.[57]

increases the rate of de-esterification and β-elimination of the glycosidic bonds.[9] This β-elimination includes cleavage of glycosidic bonds and leads to formation of a glycosyl anion and a double bond between C-4 and C-5 (Figure 11.4).[9] It has been reported that an increase in temperature increases the rate of β-elimination reaction relative to the rate of de-esterification, causing extensive main-chain breakdown.[54–56]

11.8.2 Alkali De-Esterification

Most studies use sodium hydroxide (NaOH) for the alkali de-esterification of pectin. Alkali de-esterification has a higher rate of reaction than acid de-esterification.[9] It has been reported that acid de-esterification occurs at a rate of less than 0.01% of that for the corresponding base catalyzed reaction at similar temperatures and concentration of reagents.[46] However, it is accompanied by severe de-polymerization.[46] Under alkaline conditions, or even near pH 7, de-polymerization by β-elimination occurs extensively and rapidly.[9,51] The electron-withdrawing methyl carboxy group, activates the α proton (H5) to removal by alkali. Pectate, which has a free carboxylate, is quite stable under the same conditions.[55] The degradation of pectins increases with pH, temperature and the degree of methyl-esterification (DM).[55]

Hotchkiss *et al.*[58] have reported that at pH 10 using 5.0 M NaOH and at 30 °C, the rate of β-elimination was much faster than that of methyl de-esterification. The result obtained by Hotchkiss *et al.*[58] was in agreement with that reported by Renard and Thibault.[59] Renard and Thibault[54] further suggested that for this to occur requires two adjacent methylated GalA residues.

Capillary zone electrophoresis (CZE) and nuclear magnetic resonance (NMR) studies indicated alkali de-esterified pectin yielded a random distribution of carboxylates.[58–61] In contrast, Taylor[51] suggested alkali de-esterification yielded a blockwise distribution.

11.8.3 Ammonia De-Esterification

Ammonia de-esterification yields a blockwise distribution of amide groups and a random distribution of free carboxyl groups.[51] Reagents used for amidation include ammonia, hydroxylamine and others.[62] LM pectin produced by ammonia de-esterification is known as amidated LM pectin (Figure 11.5). Renard and Thibault[54] have reported that de-esterification of pectins with ammonia proceeds at a slower rate than with sodium hydroxide at the same pH. As the time of reaction increases, the degree of esterification (DE) decreases, degree of amidation (DAm) increases and the GalA content remains constant.[54] The neutral sugar composition and side chains remain constant.[63] The molecular weights of amidated LM pectin do not change drastically as compared to the initial HM pectin used, suggesting that the reaction does not lead to drastic de-polymerization under these conditions.[64]

11.8.4 Enzymes De-Esterification

Enzyme de-esterification has becoming increasingly popular recently for producing LM pectin. It has a higher rate of reaction compared to acid and alkali de-esterification. One of the most commonly used enzymes is pectin methylesterase.[64] Enzyme de-esterification depends largely on the origin of pectin (higher plants) and the source of enzyme (microorganisms). Both the random or blockwise distribution of free carboxyl groups can be obtained.[65,66]

Pectin methylesterases in plants have been reported to produce a blockwise pattern of carboxylic groups.[62] They proceed linearly along the pectin backbone, searching for a free carboxyl group or the chain reducing end to initiate the process and produce a block of de-esterified pectin.[58] NMR spectroscopy indicated that the modes of action for the pectin methylesterases can be of single-chain or multiple-attack, depending on the pH.[58] Single-chain reaction of the enzyme acts to produce a blockwise pattern by

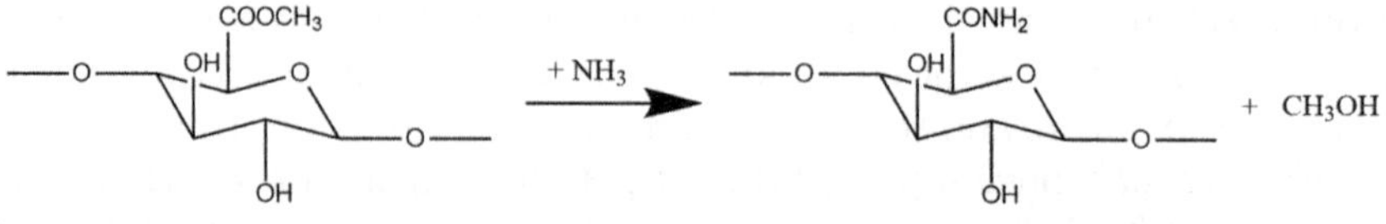

Figure 11.5 Ammonia de-esterification of pectins, yielding amidated LM pectins.

moving along a single pectin chain while multiple-attack involves multiple attachments of the enzyme, producing a shorter block of de-esterified pectin.[58,67] Pectin methylesterases from microbial and fungal sources have been reported to produce a random pattern of carboxylic groups. An example of a plant source for pectin methylesterases is Valencia orange peel. *Aspergillus niger* is the most commonly used fungal species for industrial production of pectrolytic enzymes.[58]

Lower gel strength is obtained using LM pectin produced by pectin methylesterases from plants compared to that from microbial or fungal sources. This is most likely due to the non-random distribution of methyl ester groups produced by the former. In addition, enzymatic de-esterification removes a small number of non-uronide materials.[67,68] Enzyme de-esterification is an efficient process for hydrolyzing the methyl ester groups of pectin with little chain degradation.[46] It has been suggested that molecular weight could play a key role in the surface properties and emulsion stabilizing properties of demethylated pectin.[65]

11.9 Gelling Mechanism of Pectin

Pectin acts as a gelling agent, thickening agent and as an emulsifier. The physical properties of a gel are the consequence of the formation of a three-dimensional network of polymer molecules.[69] Pectin gel is formed when portions of HGA are cross-linked to form a three-dimensional crystalline network in which water and solutes are trapped. The conditions required for gelation and the properties of the gel ultimately depend on the molecular structure, the intermolecular forces which hold the network together and the nature of the junction zones where the polymer molecules are cross-linked.[70] The junction zones in polysaccharide gels are complex and the molecular structures are held together by a large number of individually weak interactions, such as electrostatic interactions and hydrogen bonds.[69] HM and LM pectin have different gelation mechanisms, although the gel characteristics are governed by the same macromolecular properties, such as the composition, size and conformation of the polymers.[70] The degree of esterification determines emulsion formation, surface tension and gel characteristics.[71] In HM pectin, the junction zones are formed by the cross-linking of HGA by hydrogen bonds and hydrophobic forces between methoxyl groups, both promoted by high sugar concentration and low pH. In LM pectin, junction zones are formed by calcium cross-linking between free carboxylate groups.

11.9.1 Gelation of High Methoxyl Pectin

HM pectin gelation involves several kinds of intermolecular interactions. Gels of HM pectin are typically obtained in the presence of high sucrose concentrations or other similar co-solutes to achieve the necessary low water activity to minimize pectin solvent interactions and low pH to reduce ionization of the carboxyl groups thereby minimizing electrostatic repulsions.[9,75] The non-dissociated carboxylic acid groups can form inter- and intra- hydrogen bonds with secondary alcohol groups. Gels formed under these conditions are stabilized by a subtle balance of interactions, involving the formation of aggregated helices supported by hydrogen bonds and grouping of methyl-ester groups, through hydrophobic interactions, within a cage of water molecules.[14] The contribution of hydrophobic interaction is much smaller than that of hydrogen bonding, but it is essential to overcome the free energy barrier for gelation. Unlike most polysaccharide gels, the structure of HM pectin sugar gels is considered to be irreversible on heating.[9]

11.9.2 Gelation of Low Methoxyl Pectin

Although LM pectin is similar to HM pectin in terms of the formation of a continuous reticulum, it is of ionic cross-linkages *via* calcium bridges between two carboxyl groups from two different chains in close proximity instead of hydrogen bonds and hydrophobic interactions, forming a three-dimensional macromolecular network (Figure 11.6).[76] The special gelation of LM pectins is described by the 'egg-box' model.[77] This model

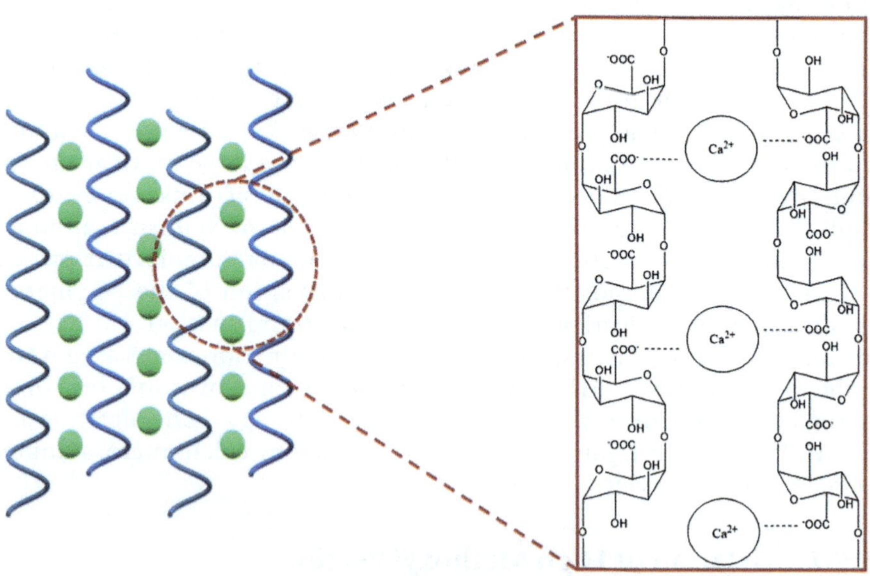

(a) (b)

Figure 11.6 Gelation of LM pectin (a) the 'egg-box' model (b) structure of junction zones in LM pectin gelation.[78]

involves junction zones formed by the ordered, side-by-side associations of galacturonans, whereby specific sequences of GalA monomers in parallel or adjacent chains are linked intermolecularly through electrostatic and ionic bonding of carboxyl groups.[78,79] The gelation mechanism of LM pectin was found to be similar to that of alginate, a polysaccharide consisting of 1 → 4 linked D-guluronate and L-mannuronate that can appear in homopolymeric blocks of each monomer.[80] It has been suggested that the egg-box model fits alginate, but it can only be an approximation for the gelation mechanism of LM pectin.[81] The latter is better described as a 'shifted' egg-box, since one of the chains is slightly shifted with respect to the other.[80] Egg-boxes formed between two neighboring chains are stabilized by van der Waals interactions and hydrogen bonds, in addition to electrostatic interactions.

A two-stage process has been suggested with an initial dimerization of the molecules and subsequent aggregation of these pre-formed dimers.[9,76] It is generally accepted that the junctions consist of dimers in 2_1 helical symmetry.[9,76,78,82] The junctions are formed between unbranched non-esterified galacturonan blocks bound together non-covalently by coordinated calcium ions. This complex involves coordination bonds utilizing the unfilled orbitals of the calcium ion.[83] The oxygen atoms of the hydroxyl groups, the ring oxygen atoms and the bridging oxygen atoms of the component sugar units participate in the bonding process through their free-electron pairs. The life of the junction depends on the strength of the electrostatic bonds. Formation of cooperative egg-boxes is only possible when stretches of subsequent non-methoxylated GalA units of minimal length are present.[80] The bonds are stable whenever there are at least six consecutive carboxyl groups on the interior of each participating chain.[80] Some studies have estimated the minimum number of successive non-methoxylated GalA residues necessary to form egg-boxes to be nine, six to thirteen, fourteen or possibly up to twenty.[50,78,80,84] The occurrence of methyl ester groups in the primary backbone limits the extent of such junction zones leading to formation of the gel. The study of the gelation of low-methoxylated pectin at pH 3 and 30% saccharose with different contents of Ca^{2+} was reported.[85] The initial gel formation process started with ionic interactions (egg-box junction zones and random cross-links) through calcium bridges and hydrophobic interactions when the temperature was above 60 °C, completing at around 40 °C. When the temperature was below 50 °C, hydrophilic interactions and inter-dimer aggregations dominated the gelation during the cooling process.

In addition to the above two well-known gelation mechanisms, some pectin can gel through formation of covalent cross-links.[80] Specifically, sugar beet pectin is known to carry ferulic acid on neutral sugar side chains which can be cross-linked by coupling reactions with oxidative enzymes such as laccase or peroxidase, resulting in the formation of a covalently linked network known as feruloylated pectin.

11.9.3 Comparison Between Gelling Mechanisms of High Methoxyl Pectin and Low Methoxyl Pectin

Comparing HM and LM gelation, the junction zones in HM pectin gels are predominantly through aggregates involving a large and variable number of pectin chains, showing no competitive inhibition by addition of short and similar chain segments, whereas the principal mode of intermolecular association in LM pectin gels is through junctions involving a small and definite number of chains.[9] The multifunctionality of pectin originates from the nature of its molecules in which there are polar and non-polar regions that enable it to be incorporated into different food systems. DM and molecular weight are also important. HM pectin is mainly employed in high sugar jam, jellies preserves and confectionery products.[14] In contrast, LM pectin is employed in low calorie and dietetic foods to prepare gels with a reduced level of dissolved solids and therefore of reduced calorific value.[86] LM pectin is also used in dairy products[14] and in biomedical applications such as PecSys® – an *in situ* nasal gelling technology[87] – and for the oral, sustained delivery of paracetamol.[88]

11.9.4 Effect of Pectin Modification on Gelling Mechanisms

Pectin formulations can be tailored into gels, 3-D matrices, films, micro- or nano-particles. However, there is a lack of reproducible performance. The obstacles include the large diversity of molecular characteristics, leading to problems in quality control and quality assurance.[89] Solutions to resolve these problems fall into two categories: the development of new technologies for pectin isolation, purification and characterization and the modification of pectin macromolecules.[90] Chemical modifications of pectin have improved the polysaccharide for several specific applications. Chen and coworkers[91] have published a thorough review on chemical modifications of pectin, which includes substitution, chain elongation and de-polymerization.[91] Alkylation, amidation and thiolation are among the modifications related to the gelling mechanism of pectin.

HM and LM pectin have different gelling mechanisms and, depending on DM, they have different commercial applications. It is increasingly apparent that the pattern of methyl-esterification is critical for determining rheological properties. The methylation pattern may influence the properties of pectin to a large extent, not only by modulating calcium binding but possibly also by influencing de-polymerization rates.

Acetylation decreases the stability of calcium binding and of LM pectin, hampering gelation to the extent that complete inhibition of gelation occurs when one out of eight D-galacturonic acid are acetylated at O-2 or O-3. This has been ascribed to a steric effect of the acetyl groups that prevents, to a certain extent, access of calcium ions to the necessary proximity of two neighboring carboxyl groups. The presence of acetyl groups can also lower the binding strength of the cation to the individual galacturonic acid residues or hinder the adoption of binding-favorable conformations by the polymer.

Amidated pectin is important in food technology with good gelling properties and reduced sensitivity towards Ca ions and pH.[80] The gel strength of amidated LM pectin is stronger than that of LM pectin.[92] The gelation of non-amidated LM pectin is reported to involve ionic interaction between GalA residues in a calcium-dependent manner, whereas amidated LM pectin involves hydrogen bonding between amidated GalA residues. Amidation, therefore, plays an important role in increasing gel strength. Furthermore, gels from amidated pectin are thermos-reversible. They can be heated and solidify again on cooling, whereas conventional pectin-gels will afterwards remain liquid.[91]

The mucoadhesive properties of natural polysaccharides are generally weak.[91] The non-covalent interaction between natural polymers and a mucus layer can be improved by thiolation, permitting the formation of covalent disulfide bonds with the cysteine-rich subdomains of mucus.[93] Thiol containing ligands can be introduced onto pectin by formation of either amide or ester bonds.[91]

11.9.5 Factors Affecting the Gelation of Pectins

The viscoelastic behavior of HM and LM pectin gels is strongly dependent on intrinsic and extrinsic factors.

11.9.5.1 Intrinsic Factors Affecting the Gelation of Pectins

11.9.5.1.1 Molecular Arrangement. The occurrence of sugar monomers such as rhamnose, whose dimensions are not compatible with the geometry of the junction zones formed by GalA monomers, impedes junction-zone formation.[76] Rhamnose occurs mainly in the hairy regions of citrus and apple pectins. These hairy regions obstruct the molecular orientation necessary for the junction zones to develop. Sugar beet exhibits the poorest gelling properties of the three commercial pectins and this can be attributed to its relatively high rhamnose content (Table 11.1).[94]

11.9.5.1.2 Molecular Weight. Gelation of pectin is influenced by its molecular weight and the rigidity of the gel is determined by the number of effective junctions formed per chain.[76] The lower the molecular weight, the less junction zones per molecule can be formed, decreasing the extent of

Table 11.1 Rhamnose content (%) of commercial pectins. Adapted from Axelos and Thibault[94]

Pectins	Rhamnose content (%)
Sugar beet	≈5.5
Apple	2.0–2.5
Citrus	≈1.5

Table 11.2 Molecular weight (dalton) of commercial pectins. It should be noted that these values will vary with extraction procedure.

Pectin	Molecular weight (dalton)
Sugar beet	35 500–44 700 [95,96]
Citrus	90 000 ± 10 000 [97]
Apple	64 000–100 000 [98]

cross-linking.[76] This will weaken the overall gel strength[23] (Table 11.2). Sugar beet pectin, with the lowest molecular weight, has the poorest gel strength among the three pectins. Accurate determination of molecular weight in pectin is difficult due to the extreme heterogeneity of pectin samples.

11.9.5.1.3 Carboxyl Distribution. Gels produced by blockwise de-esterification are weaker than those produced by random de-esterification.[59] Amidation increases the gelling ability of LM pectins.[99] Amidated pectins need less calcium to gel and are less prone to precipitation by high concentrations of calcium ions than are other forms of LM pectin.[62] However, a blockwise pattern of de-esterification leads to LM pectin that is extremely sensitive to low calcium levels.[62]

11.9.5.1.4 Degree of Esterification. The DM of pectin determines the linear charge density of the macromolecule, and thus it is the most important factor influencing the ion binding properties of pectins. Calcium cooperative binding is negligible when more than 40% of the carboxyl groups are randomly esterified.[14] Thus, gel forming ability of LM pectin increases with decreasing DM which has increased numbers of ionizable carboxylic groups.[100] These groups react with calcium ions to form the 'egg-box' model of LM pectin gel. However, it should be noted that, although pectin gels are stronger when DM is low, at very low DM pectin gels may become very brittle.[101]

The affinity of pectin for calcium ions is affected by acetylation because the size of acetylated galacturonic residues is not compatible with the severe topological constraints of the chain–chain association within the junction zones.[102] The higher the DA of the pectin molecule, the weaker the strength of the gel formed.[78] Thus, acetyl groups impart poor gelation characteristics to LM pectins.

11.9.5.2 Extrinsic Factors Affecting Gelation of Pectins

11.9.5.2.1 Ionic Strength. The interaction between polyelectrolytes and counter-ions is influenced mainly by the linear charge density of the macromolecule, expressed by the distance between adjacent charged groups. It has been shown that the intrinsic ionization constant pK_a of pectins is independent of the pectin concentration, DE, degree of amidation or degree of polymerization and has a value of around 2.9–3.2.[9] For a given pH, the degree of ionization increases with increasing pectin concentration. The apparent pK_a depends on the effective charge density of the pectin. Generally, at low ionic strength, the calcium–pectin interactions seem to be essentially cooperative whereas at increased ionic strength by, for example, the addition of sodium chloride solution, the calcium–pectin interactions are anti-cooperative in nature.[103]

11.9.5.2.2 pH. pH and ionic strength are co-related, affecting the pK_a of the pectin molecules. Interestingly, the binding of multivalent ions to pectin increases with increasing pH, showing a maximum level at pH 5.0–7.5.[9] The binding affinity of calcium to pectin decreases when the ionic strength increases. The stability constant of calcium pectinate decreases considerably with increasing ionic strength, which is attributed not only to the change of the calcium activity coefficients with ionic strength, but also to ion-exchange equilibria at the carboxyl groups of pectin.[9] Greater acidity decreases the number of dissociated carboxyl groups and therefore decreases the possibility of ionic cross-links.[103,104] The charge in junction-zone cavities is neutralized by hydrogen ions at low pH, and this causes the aggregation and precipitation of pectins.[76]

11.9.5.2.3 Temperature. LM pectin gel is thermally reversible.[105] However, higher temperature decreases the strength of the bonding forces within the junction.[106] This is an advantage for preparation of LM gels with calcium ions. Commercially, LM pectins are usually dispersed in aqueous calcium solutions at about 70 °C and then cooled slowly. The sol–gel transition temperature is influenced mostly by the amount of pectin and calcium.[103] The binding of calcium ions to pectin is less stable at high temperature and thus more calcium ions are required to form an elastically active junction zone.[103]

11.9.5.2.4 Presence of Co-Solutes. Different co-solutes affect the rheological and morphological structures of the gel formed.[103] LM pectin gels containing sodium chloride are more rigid than gels without sodium chloride. This effect is explained in terms of electrolyte behavior. In the presence of increasing concentrations of sodium chloride, the charge of the polymer is increasingly screened, allowing pectin chains to approach each other, which in turn leads to a decrease in the rate of the ionic bridging function.[103] It is possible that a slower rate of gelation will initiate a greater number of smaller junction zones and consequently will enhance the stability and mechanical strength of a gel.[103]

LM pectins are able to form gels in the presence of sugar. When soluble solids are added, an increase in the setting temperature is noticed and the gel strength also increases.[41] It has been suggested that the combined effect of pH and addition of sugars promotes gelation at lower calcium levels. Indeed, despite the decrease of the number of sequences of carboxyl groups available for calcium binding, gelation is enhanced because of the specific effect of sugars on the water activity and hydrophobic effects.[14] However, NMR spectroscopy has demonstrated that, when sugars are added, pectin and sugar molecules compete for cations.[107] Depending on the sugar's structure, a stable complex can be formed between the sugar and calcium ions. This interaction can be unfavorable to the formation of the gel, due to the decreased calcium ions available to associate with pectin and, therefore, decreased gel rigidity. The rigidity of LM pectin gels is essentially dependent on the capacity of the sugar to compete with the pectin for the calcium ions. The interaction between sugar and water is a secondary effect.

11.10 Conclusion

Pectins are versatile biopolymers, available abundantly in plants, with a diverse and complex range of structures. It is important to comprehend these complexities as these characteristics are accountable for the multiplicity of their functionalities and enable us to utilize pectins in food and nonfood applications. The sol–gel mechanisms and behaviors of pectin have been extensively studied for the last decade, however, there are still fields where our knowledge is still limited. Some of these fields include:

(i) the effect of extraction processes on structural features of pectin other than those obtained at the industrial level,
(ii) the role of each structural feature in influencing the gelling mechanism of pectins,
(iii) the gelation processes of pectins under conditions different from the two classical gelling mechanisms of LM and HM pectins,
(iv) the different specific properties imparted from different modifications of pectin and the many potential pectin modifications that have yet to be exploited.

This incoming enhanced knowledge will be important to further validate the structure and functionality of this biopolymer and expand its application uses.

References

1. B. Futrakul, M. Kanlayavattanakul and P. Krisdaphong, *Int. J. Cosmet. Sci.*, 2010, **32**, 211–215.
2. W. B. Zhang, P. Xu and H. Zhang, *Trends Food Sci. Technol.*, 2015, **44**, 258–271.
3. J. Chen, W. Liu, C. M. Liu, T. Li, R. H. Liang and S. J. Luo, *Crit. Rev. Food Sci. Nutr.*, 2015, **55**, 1684–1698.
4. J. P. Fuenzalida and F. M. Goycoolea, *Curr. Protein Pept. Sci.*, 2015, **16**, 89–99.
5. D. T. Piorkowski and D. J. McClements, *Food Hydrocolloids*, 2014, **42**, 5–41.

6. J.-H. Guo, G. W. Skinner, W. W. Harcum and P. E. Barnum, *Pharm. Sci. Technol. Today*, 1998, **1**, 254–261.
7. D. Mohnen, *Curr. Opin. Plant Biol.*, 2008, **11**, 266–277.
8. W. G. T. Willats, J. P. Knox and J. D. Mikkelsen, *Trends Food Sci. Technol.*, 2006, **17**, 97–104.
9. M. A. Rao and J. A. L. d. Silva, *Food Polysaccharides and Their Applications*, CRC Press, 2006, pp. 353–411, DOI: 10.1201/9781420015164.ch11.
10. M. S. Lima, E. P. Paiva, S. A. C. Andrade and J. A. Paixão, *Food Hydrocolloids*, 2010, **24**, 1–7.
11. Z. I. Kertesz, *J. Am. Pharm. Assoc.*, 1952, **41**, 54.
12. G. Mualikrishna and R. N. Tharanathan, *Food Chem.*, 1994, **50**, 87–89.
13. J. B. Wehr, N. W. Menzies and F. P. C. Blamey, *Food Hydrocolloids*, 2004, **18**, 375–378.
14. B. R. Thakur, R. K. Singh, A. K. Handa and D. M. A. Rao, *Crit. Rev. Food Sci. Nutr.*, 1997, **37**, 47–73.
15. W. T. Willats, L. McCartney, W. Mackie and J. P. Knox, in *Plant Cell Walls*, ed. N. C. Carpita, M. Campbell and M. Tierney, Springer, Netherlands, 2001, ch. 2, pp. 9–27, DOI: 10.1007/978-94-010-0668-2_2.
16. M. Kaya, A. G. Sousa, M.-J. Crépeau, S. O. Sørensen and M.-C. Ralet, *Ann. Bot.*, 2014, **114**, 1319–1326.
17. K. H. Caffall and D. Mohnen, *Carbohydr. Res.*, 2009, **344**, 1879–1900.
18. M. A. O'Neill, T. Ishii, P. Albersheim and A. G. Darvill, *Annu. Rev. Plant Biol.*, 2004, **55**, 109–139.
19. Y. S. Ovodov, *Russ. J. Bioorg. Chem.*, 2009, **35**, 269–284.
20. B. M. Yapo, *Carbohydr. Polym.*, 2011, **86**, 373–385.
21. F. S. Agency, *Current EU approved additives and their E Numbers*, https://www.food.gov.uk/science/additives/enumberlist#toc-5.
22. H. Kauss and W. Hassid, *J. Biol. Chem.*, 1967, **242**, 3449–3453.
23. B. M. Yapo, C. Robert, I. Etienne, B. Wathelet and M. Paquot, *Food Chem.*, 2007, **100**, 1356–1364.
24. S. M. Brejnholt, in *Food Stabilisers, Thickeners and Gelling Agents*, ed. A. Imeson, Blackwell Publishing, West Sussex, 2010, pp. 237–265.
25. M. M. Bomgardner, *Chemical & Engineering News*, American Chemical Society, Washington DC, 2013, vol. 91, p. 20.
26. C. R. Vithanage, M. J. Grimson, P. R. Wills, P. Harrison and B. G. Smith, *J. Texture Stud.*, 2010, **41**, 899–927.
27. C. D. May, *Carbohydr. Polym.*, 1990, **12**, 79–99.
28. J. Leroux, V. Langendorff, G. Schick, V. Vaishnav and J. Mazoyer, *Food Hydrocolloids*, 2003, **17**, 455–462.
29. G. Mesbahi, J. Jamalian and A. Farahnaky, *Food Hydrocolloids*, 2005, **19**, 731–738.
30. B. M. Yapo, B. Wathelet and M. Paquot, *Food Hydrocolloids*, 2007, **21**, 245–255.
31. D. D. Joye and G. A. Luzio, *Carbohydr. Polym.*, 2000, **43**, 337–342.
32. Y. Liu, J. Shi and T. A. G. Langrish, *Chem. Eng. J.*, 2006, **120**, 203–209.

33. M. L. Fishman, H. K. Chau, P. D. Hoagland and A. T. Hotchkiss, *Food Hydrocolloids*, 2006, **20**, 1170–1177.
34. M. T. Iglesias and J. E. Lozano, *J. Food Eng.*, 2004, **62**, 215–223.
35. J. Pagán, A. Ibarz, M. Llorca, A. Pagán and G. V. Barbosa-Cánovas, *Food Res. Int.*, 2001, **34**, 605–612.
36. B. M. Yapo and K. L. Koffi, *J. Agric. Food Chem.*, 2006, **54**, 2738–2744.
37. S. Levigne, M.-C. Ralet and J.-F. Thibault, *Carbohydr. Polym.*, 2002, **49**, 145–153.
38. C. F. de Oliveira, D. Giordani, P. D. Gurak, F. Cladera-Olivera and L. D. F. Marczak, *Innovative Food Sci. Emerging Technol.*, 2015, **29**, 201–208.
39. J. P. Maran, K. Swathi, P. Jeevitha, J. Jayalakshmi and G. Ashvini, *Carbohydr. Polym.*, 2015, **123**, 67–71.
40. W. J. Wang, X. B. Ma, Y. T. Xu, Y. Q. Cao, Z. M. Jiang, T. Ding, X. Q. Ye and D. H. Liu, *Food Chem.*, 2015, **178**, 106–114.
41. M. A. V. Axelos and M. Branger, *Food Hydrocolloids*, 1993, **7**, 91–102.
42. B. M. Yapo, *J. Agric. Food Chem.*, 2009, **57**, 1572–1578.
43. S.-Y. Chan and W.-S. Choo, *Food Chem.*, 2013, **141**, 3752–3758.
44. M. H. Canteri-Schemin, H. C. R. Fertonani, N. Waszczynskyj and G. Wosiacki, *Braz. Arch. Biol. Technol.*, 2005, **48**, 259–266.
45. W. J. Kim, V. N. M. Rao and C. J. B. Smit, *J. Food Sci.*, 1978, **43**, 572–575.
46. S. A. El-Nawawi and Y. A. Heikal, *Carbohydr. Polym.*, 1995, **27**, 191–195.
47. T. Duvetter, I. Fraeye, D. N. Sila, I. Verlent, C. Smout, M. Hendrickx and A. Van Loey, *J. Agric. Food Chem.*, 2006, **54**, 7825–7831.
48. T. G. Neiss, H. N. Cheng, P. J. H. Daas and H. A. Schols, *Macromol. Symp.*, 1999, **140**, 165–178.
49. I. Fraeye, A. De Roeck, T. Duvetter, I. Verlent, M. Hendrickx and A. Van Loey, *Food Chem.*, 2007, **105**, 555–563.
50. D. A. Powell, E. R. Morris, M. J. Gidley and D. A. Rees, *J. Mol. Biol.*, 1982, **155**, 517–531.
51. A. J. Taylor, *Carbohydr. Polym.*, 1982, **2**, 9–17.
52. I. N. Panchev, N. A. Kirtchev and C. G. Kratchanov, *Food Hydrocolloids*, 1994, **8**, 9–17.
53. P. G. Crandall and L. Wicker, *Chemistry and Function of Pectins*, American Chemical Society, 1986, ch. 8, vol. 310, pp. 88–102.
54. C. M. G. C. Renard and J.-F. Thibault, *Carbohydr. Res.*, 1996, **286**, 139–150.
55. J. N. BeMiller and G. V. Kumari, *Carbohydr. Res.*, 1972, **25**, 419–428.
56. P. Albersheim, H. Neukom and H. Deuel, *Arch. Biochem. Biophys.*, 1960, **90**, 46–51.
57. J. N. BeMiller, *Chemistry and Function of Pectins*, American Chemical Society, Washington DC, 1986, vol. 310, pp. 2–12.
58. A. T. Hotchkiss, B. J. Savary, R. G. Cameron, H. K. Chau, J. Brouillette, G. A. Luzio and M. L. Fishman, *J. Agric. Food Chem.*, 2002, **50**, 2931–2937.
59. C.-M. Jiang, M.-C. Wu, W.-H. Chang and H.-M. Chang, *J. Agric. Food Chem.*, 2001, **49**, 5584–5588.

60. C.-M. Jiang, S.-C. Liu, M.-C. Wu, W.-H. Chang and H.-M. Chang, *Food Chem.*, 2005, **91**, 551–555.
61. H. Grasdalen, O. Einar Bakøy and B. Larsen, *Carbohydr. Res.*, 1988, **184**, 183–191.
62. E. Racape, J. F. Thibault, J. C. E. Reitsma and W. Pilnik, *Biopolymers*, 1989, **28**, 1435–1448.
63. J. C. E. Reitsma, J. F. Thibault and W. Pilnik, *Food Hydrocolloids*, 1986, **1**, 121–127.
64. H. Anger and G. Dongowski, *Food Hydrocolloids*, 1988, **2**, 371–379.
65. Y.-H. Yoo, S. Lee, Y. Kim, K.-O. Kim, Y.-S. Kim and S.-H. Yoo, *Int. J. Biol. Macromol.*, 2009, **45**, 226–230.
66. Y. Kim, Y.-S. Kim, S.-H. Yoo and K.-O. Kim, *Food Chem.*, 2010, **123**, 451–455.
67. G. Limberg, R. Körner, H. C. Buchholt, T. M. I. E. Christensen, P. Roepstorff and J. D. Mikkelsen, *Carbohydr. Res.*, 2000, **327**, 293–307.
68. R. S. Jayani, S. Saxena and R. Gupta, *Process Biochem.*, 2005, **40**, 2931–2944.
69. D. Oakenfull and A. Scott, *J. Food Sci.*, 1984, **49**, 1093–1098.
70. M. A. V. Axelos and J. Thibault, in *The Chemistry and Technology of Pectin*, ed. R. H. Walter, Academic Press, California, 1991, pp. 109–118.
71. J. Müller-Maatsch, A. Caligiani, T. Tedeschi, K. Elst and S. Sforza, *J. Agric. Food Chem.*, 2014, **62**, 9081–9087.
72. C. Chung and D. J. McClements, *J. Food Eng.*, 2015, **158**, 13–21.
73. L. C. Vriesmann and C. L. O. Petkowicz, *Food Hydrocolloids*, 2013, **33**, 58–65.
74. F. S. Hashemi, S. M. T. Gharibzahedi and H. Hamishehkar, *RSC Adv.*, 2015, **5**, 42346–42353.
75. D. Galkowska, M. Dlugosz and L. Juszczak, *Starch/Staerke*, 2013, **65**, 499–508.
76. M. A. V. Axelos and J. F. Thibault, in *The Chemistry and Technology of Pectin*, ed. R. H. Walter, Academic Press, San Diego, 1991, pp. 109–118, DOI: 10.1016/B978-0-08-092644-5.50011-X.
77. M. Vandenbossche, M. Jimenez, M. Casetta and M. Traisnel, *Crit. Rev. Environ. Sci. Technol.*, 2015, **45**, 1644–1704.
78. I. Braccini and S. Pérez, *Biomacromolecules*, 2001, **2**, 1089–1096.
79. A. Assifaoui, A. Lerbret, H. T. D. Uyen, F. Neiers, O. Chambin, C. Loupiac and F. Cousin, *Soft Matter*, 2015, **11**, 551–560.
80. I. Fraeye, T. Duvetter, E. Doungla, A. Van Loey and M. Hendrickx, *Trends Food Sci. Technol.*, 2010, **21**, 219–228.
81. P. Jantrawut, A. Assifaoui and O. Chambin, *Carbohydr. Polym.*, 2013, **97**, 335–342.
82. O. A. Patova, V. V. Golovchenko and Y. S. Ovodov, *Russ. Chem. Bull.*, 2014, **63**, 1901–1924.
83. I. Ventura, J. Jammal and H. Bianco-Peled, *Carbohydr. Polym.*, 2013, **97**, 650–658.

84. I. Fraeye, E. Doungla, T. Duvetter, P. Moldenaers, A. Van Loey and M. Hendrickx, *Food Hydrocolloids*, 2009, **23**, 2069–2077.
85. H. Kastner, U. Einhorn-Stoll and B. Senge, *Food Hydrocolloids*, 2012, **27**, 42–49.
86. H. Liu, X. M. Xu and S. D. Guo, *LWT–Food Sci. Technol.*, 2007, **40**, 946–954.
87. J. Castile, Y.-H. Cheng, B. Simmons, M. Perelman, A. Smith and P. Watts, *Drug Dev. Ind. Pharm.*, 2013, **39**, 816–824.
88. S. Miyazaki, H. Murofushi, T. Shimoyama, K. Itoh, M. Kobayashi and D. Attwood, *Pharm. Dev. Technol.*, 2013, **18**, 1259–1264.
89. E. A. Günter, O. V. Popeyko, P. A. Markov, E. A. Martinson, S. G. Litvinets, E. A. Durnev, S. V. Popov and Y. S. Ovodov, *Carbohydr. Polym.*, 2014, **103**, 550–557.
90. L. Liu, M. Fishman and K. Hicks, *Cellulose*, 2007, **14**, 15–24.
91. J. Chen, W. Liu, C.-M. Liu, T. Li, R.-H. Liang and S.-J. Luo, *Crit. Rev. Food Sci. Nutr.*, 2014, **55**, 1684–1698.
92. C. Tamura, M. Nakauma, H. Furusawa, T. Kadota, Y. Kamata, M. Nishijima, S. Itoh and Y. Sugita-Konishi, *Carbohydr. Polym.*, 2013, **93**, 747–752.
93. F. Hintzen, S. Hauptstein, G. Perera and A. Bernkop-Schnürch, *Eur. J. Pharm. Biopharm.*, 2013, **85**, 1266–1273.
94. M. A. V. Axelos and J. F. Thibault, *Int. J. Biol. Macromol.*, 1991, **13**, 77–82.
95. L. Phatak, K. C. Chang and G. Brown, *J. Food Sci.*, 1988, **53**, 830–833.
96. F. M. Rombouts and J.-F. Thibault, *Carbohydr. Res.*, 1986, **154**, 177–187.
97. S. E. Harding, G. Berth, A. Ball, J. R. Mitchell and J. G. de la Torre, *Carbohydr. Polym.*, 1991, **16**, 1–15.
98. C. M. G. C. Renard, A. G. J. Voragen, J. F. Thibault and W. Pilnik, *Carbohydr. Polym.*, 1990, **12**, 9–25.
99. F. Capel, T. Nicolai, D. Durand, P. Boulenguer and V. Langendorff, *Food Hydrocolloids*, 2006, **20**, 901–907.
100. J. F. Thibault and M. Rinaudo, *Biopolymers*, 1986, **25**, 455–468.
101. I. Fraeye, I. Colle, E. Vandevenne, T. Duvetter, S. Van Buggenhout, P. Moldenaers, A. Van Loey and M. Hendrickx, *Innovative Food Sci. Emerging Technol.*, 2010, **11**, 401–409.
102. C. M. G. C. Renard and M. C. Jarvis, *Carbohydr. Polym.*, 1999, **39**, 209–216.
103. C. Garnier, M. A. V. Axelos and J.-F. Thibault, *Carbohydr. Res.*, 1993, **240**, 219–232.
104. S. M. Cardoso, M. A. Coimbra and J. A. Lopes da Silva, *Food Hydrocolloids*, 2003, **17**, 801–807.
105. P. M. Gilsenan, R. K. Richardson and E. R. Morris, *Carbohydr. Polym.*, 2000, **41**, 339–349.
106. D. Lootens, F. Capel, D. Durand, T. Nicolai, P. Boulenguer and V. Langendorff, *Food Hydrocolloids*, 2003, **17**, 237–244.
107. C. R. F. Grosso, P. A. Bobbio and C. Airoldi, *Carbohydr. Polym.*, 2000, **41**, 421–424.

CHAPTER 12

Perspectives on the Development of the Personal Care Industry

XIAN JUN LOH*[a,b]

[a]Institute of Materials Research and Engineering, A*STAR (Agency for Science, Technology and Research), 2 Fusionopolis Way, Innovis, #08-03, Singapore 138634, Singapore; [b]Department of Materials Science and Engineering, National University of Singapore, Singapore 117574, Singapore
*E-mail: lohxj@imre.a-star.edu.sg

12.1 Introduction

The cosmetics and personal care industry has endured several years of anaemic demand from consumers due to a prolonged global economic downturn and high levels of unemployment in the US and Europe. The industry has seen a steady stream of consolidation and today has three major players—L'Oreal, Procter & Gamble and Unilever. Companies are consolidating and merging product lines as well as dumping products that lag in performance. The drawback to all this is that there is an excess capacity of chemists from these companies who lost their jobs due to these acquisitions. The sudden flux of chemists represents an opportunity for other (smaller) companies to hire these talents and enhance their research and development capabilities. Consumers typically keep a tight lid on their spending on cosmetic and

RSC Polymer Chemistry Series No. 20
Polymers for Personal Care Products and Cosmetics
Edited by Xian Jun Loh

Published by the Royal Society of Chemistry, www.rsc.org

personal care products, especially the "under 30" demographic. This group is typically a major consumer category for these products and the hardest hit by the weak labour market. Overall, amid widespread belt tightening, many consumers have traded down from luxury brands to lower-cost alternatives, including mass market and private label products, a sector that now actually seems like one of the few growth spots in the industry.

12.1.1 The Rise of Testing Laboratories

Another bright spot is the cosmetics and personal care laboratory sector, especially the independent contract segment. Economics is a big part of the reason cosmetic companies outsource to contract labs. Companies want to keep their costs low but their quality and standards high enough to pass Food and Drug Administration (FDA) and other regulatory and quality hurdles. In this tough economy, the larger cosmetic companies try to shed the overheads and other costs linked to their in-house laboratories by outsourcing more work to contract labs. There are also practical and technical reasons for outsourcing laboratory work. Customer relationships are mostly long term and contract labs become highly familiar with customers' products, preservatives and operational systems, making it easier for them to participate in such in-house staples as research and development (R&D) and safety testing before and after products enter the market. Furthermore, preservative efficacy testing entails inoculating and testing products against bacterial and fungal (yeast and mould) contaminants to ensure potency. It is not ideal for large cosmetic companies to have live bacteria in their facilities. In addition, they also want to have data from credible independent labs for the FDA. These companies work with the internal teams of cosmetic companies and bring new knowledge into their organisations. Many companies go to universities and outside consultants to bring new light to the problems that they are trying to solve or new products they are trying to develop. The problem of this approach is that companies sometimes avoid hiring people on a long-term basis because they feel they could get things done outside for relatively lower cost. Whether or not this is the case, many companies are also very careful when developing new products or intellectual property, because they do not want the new knowledge they generate to go outside their organisations. While the large corporations have their own testing facilities, they do not have the capacity to do so on a timely basis. Companies reach out to contract labs for unbiased opinions. These labs are able to compare their products, using double-blind testing, with those of competitors and provide an independent opinion. Companies could struggle to carry out such testing in-house due to a lack of intellectual capability and various other factors, such as their limited usage of certain capabilities, the unavailability of specialised staff and high costs. Under such circumstances, independent labs represent a much better alternative for them.

An example of a type of laboratory that provides such services is Q Laboratories. Its worldwide client roster includes 13 Fortune 500 companies, and its services include testing for contaminants, shelf life analyses, verification of raw material and active ingredient claims, and quality control. A significant part of the lab's work is with raw material companies during their ingredient development stage. Ingredients are rigorously tested to establish their efficacy, safety and potential shelf life. Raw material producers need such data to prove the viability of their ingredients in a cosmetic formulation and, for new ingredients, to demonstrate how their application could lead to product improvements. Developers of raw materials use systematic and detailed laboratory investigations to develop, discover, prepare and test new cosmetic ingredients. Usually, tests outside of the R&D and QA/QC sphere are generally farmed out to contract labs. Repeated patch testing to show safety for use on the skin, *in vitro* eye testing, chemical testing and organoleptic evaluations are done at in-house facilities, while chemical tests, claims substantiation and eye irritation tests, among others, are done at contract labs. Part of the contract labs' responsibility entails evaluating and providing data on any safety and toxicity issues attributable to the ingredients and supplying guidance for their proper handling—typically in a material safety data document, which is available both to the industry and to consumers. A Certificate of Analysis (C of A) also accompanies raw materials. The certificate lists key attributes of the ingredient, including its molecular weight, pH, colour, odour, texture, viscosity, boiling and melting points, and microbial content. Suppliers also maintain records of key analyses—MS-GC, HPLC and FTIR.

Materials for personal care and home care have to fulfil many different requirements regarding performance, aesthetics, costs and safety. This has consequences for the use of new technologies for cosmetic applications. Even if the performance is sufficient or far surpasses expectations, if the other parameters do not meet the requirements, the new technology will still not be applied. There are challenges in applying polymers in cosmetic products. Although there are many materials available in scientific literature, only a small subset has been approved for cosmetics. While there have been numerous basic studies with polymers which show superiority in certain aspects of performance, more could still be done. The reasons for this are manifold and depend on the specific application. From an economic point of view, a high polymer concentration is one aspect which raises the cost of the cosmetic product. It is well known that most cosmetic products have water as their main ingredient. Furthermore, the formulation of a polymer depends on various, even more important, parameters such as the type of oil, the chosen surfactant, the temperature and the electrolyte content, to mention just a few. Phase stability is not usually very easily achieved. The formulation of polymers can be controlled by adjusting the temperature, the electrolyte content or the hydrophilic–lipophilic balance by varying the ratio of different surface active agents. All these techniques are applicable in cosmetics and have specific advantages and disadvantages. For example, most consumer products require a temperature stability which is usually not achieved with temperature-induced microemulsions. Therefore, no general rule exists for the most suitable type of polymer material in cosmetics.

Once the safety of the ingredient is ascertained, the development work on the ingredient begins. During development work in the laboratory, the challenge is to create a stable product that delivers the desired performance.

Once a product candidate is considered viable, there is a scale-up process that includes the creation of a pilot batch in a production setting that incorporates stability testing at elevated temperatures and controlled humidity. This is followed by microbial challenge or preservative effectiveness testing to determine how well a product can stand up to microbial insult. If it is evaluated positively at this stage, it moves on to production. Then quality assurance is performed, as well as organoleptic testing; *i.e.*, the touch and feel of the product. Additional testing at this stage includes viscosity analysis, pH and further analysis of the pivotal ingredients. These laboratories are regulated by the FDA and they undergo audits and registration with the FDA for product testing. In the US, the FDA is responsible for regulating personal care products. Cosmetic companies are legally bound to prove safety and ensure that their products are not misbranded nor adulterated with foreign matter or improper ingredients. In situations where this happens, the FDA has the authority to inform the public and initiate a recall of the product. There is a fine line separating over-the-counter (OTC) drugs and cosmetics in labelling claims in the US. Cosmetics claims are regulated by the FDA, and this is especially important for foreign firms trying to export to the US. The Food and Drug Cosmetic Act defines cosmetics by intended uses such as cleansing, beautifying and altering appearance, all of which are cosmetic claims. Claims, exaggerations and poetic language, like "energizes your hair," cannot be regulated by the FDA because they are not quantitative. Cosmetic producers are also allowed to use language with claims like "covers up fine lines," however, they cannot say "renews the skin" because this implies that the product affects the structure of the skin, which is a drug claim and implies effects on the function of the body. However, review times are increasing, and the cost is increasing because the FDA now charges for this process. There will be increased technical capability in the industry because instrumentation is becoming less expensive and more available to midsize companies. There will be an increased emphasis on international products, and the movement of products into a more global marketplace, which will make regulations much more important.

12.1.2 Training a Workforce

Despite the extensive use of scientific instrumentation and advanced analytical tools, the development of cosmetic and personal care products is half science and half art. There needs to be considerable reliance on additional expertise regarding fragrance, texture and feel on the skin, among other specialties. Cosmetic formulators are typically chemists, and the most common skill sets include chemistry and microbiology. The development of cosmetic science degrees and diplomas is a direction towards training a workforce for this industry sector. Typically, the institutions provide training in chemistry with applications in perfumery and cosmetic science. Due to the interdisciplinary nature of the industry, students are exposed to an integrated learning experience where they build on a strong chemistry foundation (half science) and apply their knowledge in specific applications using their senses (half art). This training leads to specialisations in chemistry, perfumery and cosmetic science, and students are exposed to collaborations with industry partners to commercialise the student formulations. They also get to experience internship with perfumers, chemists, product formulators or dermatologists at chemical or cosmetic companies or fragrance and flavour houses. There are literally thousands of scientists and chemists working in the cosmetic industry. There are various types of companies that employ cosmetic scientists and chemists. These include finished goods manufacturers, contract manufacturers, raw material suppliers and testing laboratories. Some of the possible jobs include cosmetic formulators, where the primary role is to invent and create new formulations. Most of these jobs are with finished goods and contract manufacturers although a few raw material suppliers employ formulators in their technical services departments. The cosmetic industry requires quality assurance and a quality control chemist (QA/QC) is in high demand in virtually all of the companies. For people who prefer work that is closer to science, analytical services offer that opportunity. Most raw material suppliers and finished goods manufacturers have analytical departments. There are openings in process engineering (PE). These jobs focus on building things and engineering and are applicable to almost any cosmetic company with manufacturing facilities. Finally, a synthetic chemist is important to raw material suppliers; polymers and small molecules are made and designed by these specialists who are very much sought after for their creativity and skills.

12.2 Conclusion

From the chapters presented in this book, one might gain the impression that most of the work in polymers has been done and completed over the years and that the application of polymers in the personal care industry requires just a heavy dose of innovation. In this book, I sought to look at the applications that are products of our innovation but have various difficulties being put into practice. Some of the ideas might never be successful but their tremendous potential makes the risk worth taking. This book aims to capture the most exciting scientific ideas in the personal care sector and to converge upon the realities of the field. I have also presented certain topics which are related to medical technologies and are equally applicable to cosmetics. The topics covered in this book do not cover the whole range of future challenging projects that needs to be addressed. Exciting work such as materials that are triggered with non-invasive external stimuli, such as electromagnetic fields or simply light, could well be the future materials for cosmetics. These could be superior to changes in composition, temperature or pressure. On the basis of these promising new concepts, additional major breakthroughs in the cosmetic industry can be expected in the future.

Subject Index

References to tables and charts are in **bold** type